Contents

Acknowledgements VII

1 | Introduction 1
2 | On the Fringe of Empires 33
3 | Learning to Be Humanitarian Subjects 53
4 | Behind the Performance 83
5 | On the Threshold of Australia 113
6 | Domestic Humanitarianism 133
7 | Sanitizing Otherness, Becoming Australian 165
Conclusion: *Humanitarian Gestures* 205

References 217
Index 239

Acknowledgements

THIS BOOK WOULD NOT HAVE BEEN POSSIBLE without Dr. Patrick Guinness's steadfast guidance. His faith, not only that this research would be completed but his insistence that it could be improved, strengthened, and honed, saw me through many days. He consistently challenged me to critically analyze many things that would have been overlooked in my exuberance for fieldwork. Further, he patiently allowed a toddler to run wild in his office when I needed to meet but could not arrange child care. Thank you for guiding me through the ethnographic process. Dr. Francesca Merlan and Dr. Sverre Molland provided guidance at crucial junctures of this work. Dr. William Bosworth provided feedback from a discipline outside of anthropology and, in doing so, enhanced my approach to the material. A special thank you to Dr. Christina Kenny for sharing her energy and passion about gender relations. Dr. Micheal Savvas's and Dr. Marshall Johnston's careful reading and thoughtful feedback was greatly appreciated. This work has been tremendously improved by two anonymous reviewers who pushed me to think more deeply about specific interactions in the field, and to be more forthright in the discussion of how fieldwork is conducted: I am very thankful for that. Mat Buntin at University of Alberta Press has been exceptionally gracious and supportive. I could not ask for a better editor to work with. This book has benefitted from support, conversations, workshops, and presentations at the Australian National University, the Australian Anthropological Association, the

East-West Center, Fresno Pacific University, and the University of Newcastle. Sections of this work have appeared in *The Journal of Refugee Studies* and are reprinted here with permission. Of course, any errors are purely my own.

The people that shared their time, stories, and experiences with me both in Nepal and Australia are too numerous to list. I am humbled you would share your lives with me and am forever indebted to you.

Two incredible women helped me care for my children over the many years it took to research and write this book—it never would have been completed without support from Sue Nickson and Marilyn Couzens. My husband, Ray, is undoubtedly my most devoted supporter. Thank you for your love, patience, and willingness to listen to my wandering thoughts. This book is dedicated to our darling daughters: Cynthia, Heidi, and our angel Edna. I hope this work contributes by a small measure to you inheriting a more equitable world.

1
Introduction

"*Okay, now we will practise our English out loud. You, you go first.*"
The coordinator nods towards a Bhutanese refugee taking part in the community-run course in Adelaide, South Australia, on Saturday mornings.
  With assurance, she states, "*My name is Gita. I am from Bhutan.*"
The coordinator nods. "*Good, good. Next.*"
  The next man says, "*My name is Hari. I am from Bhutan.*"
  Again, the response is affirmative. The next lady begins timidly, "*My name is Yasoda, I am from Nepal…no Bhutan, Nepal.*"
  The coordinator cuts in, "*You are from Bhutan. Bhutan is your country.*" He glances sheepishly at me, writing notes, and explains, "*You know, it's a little confusing for them. Many people are illiterate.*"
  She repeats, "*My name is Yasoda. I am from Bhutan.*"
  "*Good, good,*" the leader replies. The exercise continues across the thirty participants. *My name is…I am from Bhutan. My name is… I am from Bhutan.*

SUPERFICIALLY, the above exchange suggests a solid success—an ethnic leader facilitates an English class for his community while coaching the group to provide correct answers regarding its identity in Australia. The coordinator is astute: Yasoda's wrong answer is problematic, and her answer could elicit suspicion—if her nation is *not* Bhutan, then, theoretically, she is not a refugee and does not have a claim to be in Australia. The coordinator smoothly shifts away suspicion through a sympathetic appeal—Yasoda is confused and illiterate. She lacks the skills to present herself accurately or appropriately to those outside her community. Yet it is imperative that she learns the expected response. This exercise takes on a performative dimension, which I witness as an outside researcher. The coordinator represents Yasoda to me as the figure of an ideal refugee: helpless, deserving of sympathy, and immersed in her community. The representation is strategic. Bhutanese refugee elites largely create and effectively regulate the image of a Bhutanese refugee community to maintain humanitarian support. However, the contemporary set of expected refugee behaviours is not always the image all Bhutanese want to present. These expectations are simultaneously performances, contestations, and negotiations. This interplay between teacher and student is not an attempt to manipulate others. Rather, it is a response to contradictory and confining expectations: they are engaging in contextualized behaviours (Goffman 1959). This seemingly benign interaction is, in reality, illustrative of a multitude of tensions underlying not only this particular group's experiences but also the broader humanitarian framework.

This book is about Bhutanese who fled Bhutan, resided in camps in Nepal, and finally settled in Australia. While living in Nepal's refugee camps, the Bhutanese were described by the United Nations High Commissioner for Refugees (UNHCR) as model refugees (Ashton 1996; Human Rights Watch 2003). These model refugees are part of the UNHCR's much lauded and largest resettlement program to date (International Organization for Migration [IOM] 2015a). Government representatives, council workers, and service providers described the Bhutanese as a successful refugee community once resettled. The Bhutanese-cultivated image of ideal, model, or successful refugees speaks to their ability to reflect the values of their respective audience. It also appears to have been rewarded; globally, they are the elite few that will resettle in the West. In a world that is increasingly inequitable, the Bhutanese are the realization of "the fantasy of a global moral community" (Fassin 2012, xii). Yet this celebration of the model refugee masks as much as it reveals. Focusing on their current exemplary behaviour minimizes events that led to their exile from Bhutan and obscures power disparities embedded in the camp systems. The refugees I worked with did not view support from the UNHCR as a legal right due to their status as refugees. Rather, support was viewed as contingent on them being "deserving" or "good." In Australia, resettling refugees from UNHCR-run camps legitimizes an increasingly restrictive migration policy. Further, inequality punctuates relationships not only between humanitarian representatives both in Nepal and Australia but also among the Bhutanese as elites vie for power. Across these multiple spheres of power, certain norms, values, and groups of people are deemed acceptable, while those outside this realm are marginalized.

A common saying among the Bhutanese refugees was, "The elephant has two sets of teeth. The set they use to chew, and the set they show the world." While an elephant's tusks are impressive and play crucial social roles, the hidden teeth are what ultimately sustain the animal. This statement was used to illustrate the tension between how people present themselves and their underlying intentions or actions: the public performance that is maintained through private mechanisms. It suggests a duality that cuts to the heart of this book: compassion and repression, solidarity and discord. The tension between different moral values results in a series of performances—wealthy countries funnel billions of dollars into camps to promote the ideals of human equality while also promoting policies to close borders against "undesirables." In camps, the creation of a public figure who is acceptable within a humanitarian framework means deeply held values and cultural norms must be disavowed or hidden away. Leading up to resettlement, the Bhutanese imagined themselves as capable, future citizens. They understood they would face challenges in terms of language and social integration, but, ultimately, resettlement would mean they would no longer be refugees. In Australia, the Bhutanese found themselves again subjects of a form of "domestic humanitarianism." Their behaviour and cultural norms were critiqued in relation to an elusive multicultural ideal overlaid with expectations of "acceptable" refugee attributes. Crucially, both the camps and the suburbs of Australia should be thought of as humanitarian spaces. They are two sides of the same coin, highlighting the sovereign's ability to decide who is allowed to cross geopolitical boundaries, and which roles they will be allowed to inhabit once resettled. In both contexts, recognition is afforded, but it is fickle and tends to refract the inequality within nation-states rather than address them.

Humanitarian configurations are, despite appearances to the contrary, intrinsically linked to the nation-state. In turn, this underscores the role humanitarian ideals have in the contemporary global order.

A Changing Approach to Refugees

Though refugees are not a distinctly modern phenomenon, the post–World War II era marked the institutionalization of refugee status with the creation of the UNHCR. The 1951 United Nations Convention Relating to the Status of Refugees stipulates a *refugee* as someone who,

> owing to well-founded fear of being persecuted for reasons of race, religion, nationality, membership of a particular social group or political opinion, is outside the country of his nationality and is unable or, owing to such fear, is unwilling to avail himself of the protection of that country; or who, not having a nationality and being outside the country of his former habitual residence as a result of such events, is unable or, owing to such fear, is unwilling to return to it.

For the past six decades, this definition has remained virtually unchanged, yet the way refugees are imagined is drastically different. Until the end of the Cold War, the popular image of a refugee was politically engaged and European: capable and desirable. Refugees were imagined as heroically leaving oppressive countries to share in democratic ideals. Before the end of the Cold War, refugees acted as valuable political and ideological tools: "the figure of the refugee here is inescapably political and has a particular ideological value: the refugee was said to be 'voting with his feet' by fleeing to the west" (Johnson 2011, 1020). They validated the capitalist-democratic value system. The global political environment and the assumption that refugees brought tangible skills marked earlier refugees as respected (perhaps even coveted) prizes in the Cold War. These political and popular conceptions, in turn, influenced the institution designed to protect refugees.

The UNHCR's original role was to provide legal protection for those returning to their homelands or escaping communist regimes—essentially safeguarding Europeans moving across Europe (Long 1993). The goal of protection sought to replace what had been lost prior to or during flight—the relationship between citizen and state. Namely: the right to work, freedom of movement, the protection of human rights, and physical security (Edwards 2005). Legal protection also functioned to prevent non-refoulement or forcible removal from the country of sanctuary. During the Cold War, protection became equated with resettlement, and repatriating refugees was considered taboo (Gatrell 2013). At a political level, this resulted in "policy-makers and officials devoting much of their efforts to resettling refugees from communism" (Gatrell 2013, 7). Until the 1980s, the vast majority of refugees resettled in the United States came from the Eastern bloc (Barnett 2002). Thus, for a substantial period of time, legal protection for political refugees was central and achieved vis-à-vis resettlement. This illustrates that a particular understanding of who refugees are and what they represent is crucial in terms of the support and reception they receive. These understandings are far from static, rather they respond to shifting global contexts and changing institutional orientations.

Initially, the UNHCR focused on refugees from Europe, but events occurring in Africa and Asia became influential between the 1960s and 1980s (Loescher et al. 2008). The decolonization of Africa resulted in an expanded definition of what constitutes a refugee. The 1967 Protocol universalized refugee status beyond the boundaries of Europe. The move from assisting primarily European refugees to refugees in the broader world "was a significant turning point in the Office's geographical scope and function" (Loescher et al. 2008, 25). As the UNHCR extended its reach into Africa, the support refugees received similarly increased. No longer primarily concerned with legal protection, the UNHCR now provides material assistance to refugees. Additionally, the rapid onset of refugee-generating events made non-European refugees appear to be masses of victims requiring urgent intervention to save lives and relieve suffering. Death, disease, and misery, rather than political persecution, increasingly provided the onus for UNHCR intervention: these refugees required immediate humanitarian assistance.

This sense of urgency holds a pivotal role in providing a clear, morally driven path for action (Bornstein and Redfield 2011). The emphasis became saving lives in emergency situations. This sense of immediacy, though perhaps effective in generating donor support, obscures the political realities and historical processes that generate refugees. Refugees from outside of Europe become recreated or reimagined: the image of refugees primarily as victims is solidified. As Stein (1981, 327) has noted, "Refugees are helped because they are helpless." Stein observed this in the early 1980s—when there were conceivably two parallel groups of refugees. One group is the idealized, anti-communist, European refugees—individuals capable of decision making and planning. The other group, largely in Africa but also in parts of Asia, appeared to lack this sense of autonomy. Skran (1992, 5) argues, "differences between European and non-European refugees have been greatly exaggerated" to legitimize keeping refugees located in the Third World from accessing the Global North. Chimni (1998), however, quite rightly suggests the differences are real. Just as there were differences between the wealthy, intellectual refugee and the poor, illiterate refugee, differences between refugees from Europe and refugees from other parts of the world exist. The *perception* of what these differences mean is significant. Refugees from outside Europe are increasingly divorced from, rather than defined by, their political contexts. In Africa and in Asia, the UNHCR is increasingly becoming a humanitarian organization focused on alleviating the suffering of masses of poverty-stricken victims (Johnson 2014). Since the 1980s, the focus continues to shift towards alleviating victimization and suffering. This highlights that differences and similarities are emphasized or minimized to justify distinctive responses to different refugees.

The end of the Cold War led to new strategies of global governance and reconfigurations of social inequality (Thomas and Clarke 2013). Along with the great hope the triumph of democratic governance and the free market promised, the liberal notion of a unified humanity moving towards a better world held the day. Yet the global village's market-based economy appeared to increase, rather than alleviate, poverty in developing nations (Lai 2000). The development aid that formed a key aspect of Cold War policy seemed to have little positive impact on economic expansion (Cornwall and Brock 2005). An increasing awareness of global inequality paralleled the "introduction of a new dimension to the responsibility of the global community to take a direct hand in the resolution of...unacceptable human suffering" (Munro 1999, 466). It quickly became evident that not all suffering could be alleviated.

The contemporary imagining of refugees emerges at this juncture. Global shifts triggered firstly an international awareness of their presence as "suffering masses of humanity" that existed previously but had been largely invisible (Harrell-Bond 1986; Malkki 1995a). Their emergence onto the global stage corresponded with media campaigns by organizations such as the UNHCR (Cottle and Nolan 2007), paired with increasing celebrity promotion (Kapoor 2013; Müller 2013). These campaigns highlighted the helplessness of refugees—eliciting pity and compassion (Malkki 1996; Rajaram 2002; Höijer 2004). Secondly, refugees became synonymous with being worthy of assistance precisely because of their "bare humanity" (Agamben 1998). The refugee condition became "the most privileged amongst many inferior statuses" (Zetter 2007, 189). These changes gave rise to a global understanding of refugees that was increasingly removed from broader political processes. The 1988 World Refugee Survey illustrates this transformation:

Refugees and displaced people are victims. A few are controversial. But the bulk are regular people caught up in persecution and violence (US Committee for Refugees and Immigrants 1988, 41).

This statement raises several key themes illustrative of broader shifts. First, refugees are primarily victims. This is not entirely new; Europeans after World War II could also be considered victims. However, victimization was becoming central in defining a refugee.

The reference to a few "controversial" ones is vague, though during the late 1980s the emergence of the third image of the "refugee warriors" was deemed problematic both politically and institutionally (Zolberg et al. 1989). Refugee warriors lived in camps but also used the camps as bases to launch attacks against the regime in power within their homeland (Adelman 1998). Simultaneously dependent on humanitarian assistance to survive while engaging in an armed political struggle, they represented the emerging conflict between refugee status and political activism (Zolberg et al. 1989). The final statement, "the bulk are regular people caught up in persecution and violence," focuses on passivity. Rather than political actors, refugees are "caught up" in a situation. The situation happens to them, the "bulk" of them are reacting to events they have little to do with. This perception has some truth in it: widespread political violence impacts even those not directly involved. What is striking is the fundamentally altered perception of the refugee.

No longer perceived as fleeing due to political events or views, refugees are fleeing generalized persecution and violence. Rather than political engagement being a central, defining aspect of how a refugee was conceptualized, political involvement became equated with a controversial fringe. This lack of political motivation is normalized into a "myth of difference" (Chimni 1998). Chimni (1998) argues these differences result in a paradigm shift in responding to refugee populations and understanding the contexts that result in exile. This shift obscures the rights of refugees to seek asylum, legitimizing increasingly restrictive policies.

By the 1990s, refugees were no longer celebrated for voting with their feet, marking a definitive break in the construction of refugees in the Western imagination (Hyndman 2000). Increasingly, efforts were undertaken by the UNHCR to address the root causes of refugee movements. The earlier solution of legal protection through resettlement shifted towards the aspiration to "transform the structural conditions that make populations vulnerable" (Barnett 2005, 733). The root causes were conceived as arising from internal policies rather than larger geopolitical and economic realities (Suhrke and Zolberg 1989). This had marked consequences. Primarily, as Chimni (1998, 2000) argues, if the refugee-generating state is solely responsible for refugee flows, other countries can effectively skirt their obligation to resettle. This understanding simplifies complex global interactions, such as the impact of international sanctions, imbalanced trade agreements, colonialism, and the destruction of arable land for economic benefit (Thomas and Clarke 2013). In doing so, the international dimensions of refugee flows are effectively obfuscated. Additionally, by conceptualizing root causes as domestic issues, camps become sites to fix domestic problems via development activities. Material assistance and, increasingly, development projects replace earlier resettlement-centred approaches to refugees (Harrell-Bond 1986). The UNHCR gradually has become linked with the goal of developing refugee communities (Loescher et al. 2008). This has problematic repercussions: colouring how refugees are understood and the legal pathways that may (or may not) be available to them. Hence, "the idea of linking relief and development has serious implications for working in protracted political crises" (Duffield 2007, 189). Refugees, rather than being resettled, are now perceived as a problem requiring alternative solutions: containment and repatriation (Chimni 2009) and development. Problematically, as the following chapters illustrate, this can become a means to depoliticize the social processes that led to exile and long-term containment. These processes are not necessarily linked to the political reforms of the refugee-generating country, or a challenge to state policies that strictly regulate their borders. Rather, the problem becomes reflected onto the refugees themselves: they need to be developed.

Managing States of Exclusion

While "humanitarianism is about symptoms not causes" (Barnett 2018, 332), by the late 1980s, the UNHCR high commissioner proposed the organization needed to be confronting the root causes of refugees' exile to overcome perceived cycles of aid dependency. The root causes became linked to a lack of development. In 1987, the UNHCR and United Nations Development Programme (UNDP) began a co-operative relationship to undertake development activities with refugees (UNDP 1987). These two organizations became co-mingled to promote the harmonization of humanitarian and development action (UNDP 1987). This marriage was codified in 1997 to "promote an early and smooth phase out of humanitarian assistance in favour of sustainable local development" (UNHCR 1997, para. 8). This is an interesting shift; by the late 1990s, development programs were being criticized as failing to make the dramatic transformations once imagined (Escobar 1995; Friedman 1992). While a complete history of development theory and practice is beyond the scope of this book (see instead Rist 2014 and Stewart et al. 2018), broadly, development agendas have been criticized as a means of extending Western economic power and imperialist aspirations. It appeared to be a guise to facilitate the extraction of resources or labour, deepening global inequalities (Rist 2014). Development discourse also functioned to obfuscate the political dimension of resource allocation. In turn, this naturalized inequality through bureaucratic processes and the emphasis on technical solutions to complex social structures (Ferguson 1990). On a more abstract level, the dichotomy of developed/ underdeveloped created an enormous population that required intervention and adjustment (Shrestha 1995). This perpetuated, rather than mediated, power imbalances between "the West and the rest" (Hall

2006). In many ways, this is an extension of governance strategies used in the countries that provide funding for camps. In advanced, liberal democracies, state responsibility for social difficulties is replaced with the narrative that individuals with the right values, and who practise the correct behaviours, can overcome any issue (Rose 2017). This logic naturalizes the idea that if people are experiencing social difficulties, it is due to individual flaws rather than structural, systemic issues.

The form development took in the refugee context was presented as being distinctly *humanitarian*: morally just rather than economically corrupt. The focus was on developing refugees' understanding of democratic governance and human rights (Bakewell 2003). These were ambitious and laudable goals—beyond reproach. Yet developing populations to hold the values of universal human rights and democratic governance is still "a process whereby the lives of some people, their plans, their hopes, their imaginations, are shaped by others who frequently share neither their lifestyles, nor their hopes, nor their values" (Tucker 1999, 1). These transformations frequently required radical social changes and represented an astounding incursion into the everyday lives of refugees. Even the most intimate of relationships, say between a husband and a wife (or wives), or parents and their children, become legitimate sites of intervention. By casting refugees as a population that not only require but even benefit from development programs running in the camps (by learning the right values), their containment becomes normalized. Brohman (1995, 136) argues the "ethnocentric and ideological biases of the mainstream development framework" become obscured under scientific or, as Ferguson (1990) proposes, technical discourses. This is even more acute for development in humanitarian settings, where the moral imperative creates "a kind of protective screen whereby any specific procedure is justified" (Orford 2010, 336).

In relation to the metamorphosis of the UNHCR, Barnett (2001, 247) observed "the worrisome possibility that a more pragmatic UNHCR is potentially (though unwittingly) implicated in a system of containment." Duffield (2010) similarly cautions that the shift from protecting the insecure to developing them is not benign or inevitable. Rather, it is a means of legitimizing their management and containment (Duffield 2010). Brohman (1995, 134) argues that development programs function to "turn attention away from the basic inequalities in the international economic and geopolitical structures by placing the onus for Third World underdevelopment firmly on the South itself." Root causes are not linked to the political reforms of the refugee-generating country, or global inequities that can contribute to social unrest. Refugees, rather than larger structures and processes, become the problem due to their perceived "underdevelopment." Crucially, this legitimizes refugees' strict admission into wealthy countries. Presenting the development of refugees as morally just has unfortunate consequences. It helps to absolve moral qualms that a strict migration policy to wealthy countries raises. If refugees' interests are theoretically being advanced through a robust education in international norms and values, this minimizes the need to resettle refugees. Approaching refugees as underdeveloped normalizes the exclusion from the nation they fled or the nation they reside in.

A well-run camp, helping refugees improve themselves in line with international values, might help refugees create a semblance of normalcy, but it also inadvertently helps make the spaces morally acceptable. In many regards, refugee camps have morphed into surrogates for nations, quasi-states that international donors actively construct and reconstruct through development projects. Not only is the suffering of refugees being alleviated through the provision of material goods, but their morality is also being improved through the introduction of international norms and values. In turn, this helps legitimize refugees' containment and the institutionalization of refugee camps. These programs stabilize the national system, making the camps morally palatable to wealthy countries reluctant to consider resettlement.

In contrast to the earlier days of the UNHCR, few refugees will be resettled in the Western world (Barnett 2002). So very few refugees are admitted that FitzGerald (2019, 3) noted, "Resettlement is like winning the lottery." Resettlement is not the first solution, or even a realistic solution for refugees. Historically, refugees played a significant role in global politics—their political engagement was the crucial attribute of a refugee. This has changed. Instead of the earlier focus on the political affiliations of refugees, the philanthropy of the international community begins to take centre stage. Refugee assistance is increasingly framed as a moral act (charity/benevolence), rather than the fulfillment of either a legal obligation or a critique of the generating country's political landscape (Neikirk 2018). As political solidarity gives way, "these wandering masses in movement, certainly perceived as masses of 'victims,' but just as often as supernumerary and undesirable populations," arouse anxieties (Agier 2010, 29).

Accepting refugees for resettlement now represents a moral impulse and charitable gesture. Consequently, the political dimensions of refugees are minimized, replaced by a focus on victimization. These shifts have crucial consequences: "How we imagine particular categories of people determines how we engage with them, whom we accept as legitimate political actors and who is able to participate in our world" (Johnson 2011, 1017). Humanitarianism emerges as a central concept in both how refugees are popularly imagined and approached by institutions. This relationship now is common sense: it is a given that refugees are a humanitarian project. Limbu (2009, 267) highlights this understanding with the statement, "Refugee is first and foremost a bureaucratic and humanitarian term." This may be true, yet how and why this link is normalized, as well as its implications, merits consideration. While refugees are now imagined in a humane and sympathetic fashion, rights and obligations are becoming marginalized. This raises other questions: Which groups of refugees have become acceptable, and what mechanism is calibrating refugees? Who is most deserving? Who is allowed to enter countries where they can be safe, and why?

Humanitarianism Revealed

Humanitarianism comprises four core principles: humanity, impartiality, neutrality, and independence (Barnett and Weiss 2008). Taken together, these principles translate into the premise that all people in need are entitled to unbiased assistance, given without hope for self-benefit to the provider. Humanitarianism is, at its most basic, a commitment to helping others—a desire to alleviate suffering. Humanitarian ideals are concerned with the sacredness of human life, universal brotherhood, and relieving suffering (Ticktin 2011). The equality of people underscores this ideology and suggests humanitarian acts are (at least theoretically) free from political opportunism. These aspirations speak to humanitarianism's religious roots, a genealogy through the Enlightenment, and its successful transplant into the secular realm (Stamatov 2013). These ideals permeate our contemporary world: moral sentiments have seeped into governance, "in which particular attention is focused on suffering and misfortune," combined with a desire to alleviate them (Fassin 2012, 1). Underpinning these sentiments is a profound, yet precarious, sense of moral authority: a moral responsibility to the poor, the impoverished, the victims. This morally apt position, due to its seeming irreproachable benevolence, belies powerfully unbalanced relations. Lofty ideals obscure contradictions and paradoxes. At the foundational level, there is a striking disparity in the distribution of global wealth, stemming from historical and contemporary social arrangements. The profound disparities between the wealth of some countries and the destitution of others are not new (Milanovic 2011). What may be unique to our contemporary moment is the emerging everyday relationship between the economic "have-nots" and the "haves," due to new and traditional media. Humanitarian ideals are particularly effective because they frequently promote

a sense of gratification for having contributed to saving the suffering of the world with no challenge to their own lifestyle or global inequalities, or to the political system they are a part of...a global order that fails to engage with structural inequalities (Müller 2013, 478).

These small acts of supporting good causes create relationships that focus on alleviating suffering (Donini 2010). This is framed as an altruistic act with "a kernel of nobility in...giving money to 'good causes' to alleviate suffering" (Douzinas 2007, 13). In effect, this allows structural inequalities to go unchecked and political realties to remain unexamined. Under the premise of humanitarian generosity, the reality of vastly different economic situations and access to opportunities persists. This brings me to the second problematic aspect of humanitarianism: humanitarian endeavours are undertaken with the desire to help. Yet not everyone can be saved. Humanitarianism seeks deserving victims.

Thomas and Clarke (2013, 317) propose that the post-colonial, post-Cold War, post-9/11 world has steadily entered "a new era of moral and humanitarian protectionism," with a particular focus on victims. This category is not natural or given but a complex and morphing intersection of ideals, cultural values, and political rhetoric. Victim status has become elevated as a means for people to gain recognition and social support. In parallel to this elevation is its exclusiveness, and here emerges the thinly veiled evaluation of deserving versus undeserving. "Because giving assistance is generally regarded as charity, humanitarians also assume the power to decide *who is deserving*. Such power is highly seductive and brings out the best or the worst in us" (Harrell-Bond 2002, 68). Within this broad concept of sufferers in need of help, refugees emerge as a particular kind of victim. Examining a group that does not merit support may illuminate the specific set of variables that constitute a deserving victim in the context of refugees.

During the violent breakup of the former Yugoslavia in the 1990s, Serbia hosted thousands of refugees in dismal conditions. Despite the state's inability to provide materially for this population (reaching a million in the summer of 1999), the international image of Serbs-as-evil-aggressors translated into a lack of aid. This image was honed in the media but also reflected significant processes of valuation linked to refugee status. While refugee status is technically a right, it is being progressively treated as a privilege (Zetter 2007). This privileging creates a narrow canon of ideal refugee-victim. Serbs were perceived as politically active, the aggressors in the conflict, while other groups were perceived as helpless and hence worthy. Refugees in Serbia received less assistance and had fewer opportunities to resettle in Western countries compared to refugees from other backgrounds in the former Yugoslavia (Nikolic-Ristanovic 2003, 105). In the hierarchy of victims, they did not quite fit the image of deserving or worthy.

This links with the argument put forth in the previous section that the refugee label has, over time, been progressively moulded into an ideal-type victim devoid of political messiness. Yet supporting some groups of refugees and not others is inherently political. This illuminates a fascinating paradox in which the category of refugee maintains a distinct political charge, but to be considered a "deserving refugee," politics must be downplayed, hidden, or perceived as unproblematic. Humanitarianism thus not only seeks but constructs a particular victim. As such, despite claims of neutrality, humanitarian ideals and practices are "shaped by the political ideas of an age" (Chimni 2009, 20–21). Humanitarianism masks contradictions and valuations: who is worthy of help, and who is not. This process is problematic, for "the increased moral content of humanitarianism combined with a move towards taking sides with the victim against the aggressor runs the risk of producing a hierarchy of victims" (Fox 2001, 282).

The example of a group that were not considered deserving victims, namely Serbian refugees, can be contrasted with the construction of the ideal-type, deserving refugee. Harrell-Bond (1999, 147) found that "the documents [she] obtained from [aid] agencies emphasized images of helpless, starving masses who depend on agents of compassion to keep them alive." The UNHCR (2015) appeals to the reader: "Imagine being the mother of a sick and hungry child and having to decide between risking your life staying in a conflict or leaving behind everything in search of safety." Such documents are produced with the hope of securing funding, suggesting that helplessness and suffering are crucial hallmarks of a deserving victim; "humanitarian compassion seems increasingly reserved for those who only suffer but do not act" (Feldman 2009, 31). The theme of helplessness is particularly salient in relation to the perception of deserving refugees: they are victims uninterested in helping themselves and wholly dependent on the compassion of others. Yet their very helplessness makes their status all the more precarious: action casts doubt on their position as refugees. In turn, survivors of disaster, oppression, and persecution adopt the only persona that allows them to access support: victim. This status is tenuous; "if they do not appear 'innocent' enough, or if they otherwise do not conform to the narrative demands of this category," they risk losing support (Feldman 2009, 32).

This statement exemplifies the danger behind this shift; these understandings serve to legitimize supporting some groups, while justifying the lack of support for others. It also manifests a system of ranking that leads the beholden party to recognize that humility, rather than political action, is expected (Barnett 2012). This divorces victims from historical forces and ongoing power disparities that are the sources of contemporary circumstances (Brown 2009). As refugees are depoliticized, "inequality, subordination, marginalization, and social conflict, all which require political analysis and political solutions" become constructed as individual, natural, cultural, or religious problems (Brown 2009, 15). Obscuring these interrelated processes divorces both humanitarian governance and those governed from engaging with sources of inequality. Recognition is limited to their status as "victims," rather than the complex historic, social, and political events that led to exile. The compassionate approach to deserving refugees seems superficially to be humane, just, and inherently good. The victimization category is problematic in the context of refugees, however, because it positions them as moral categories rather than political actors. In turn, refugees are placed in a precarious position: they become dependent on the compassion of others that can be withdrawn at any time (Feldman 2009). This also limits presumed capabilities and reinforces power imbalances. Still, it is also a category that has strategic value.

In Fassin's (2012) *Humanitarian Reason*, a few pages are devoted to a group of asylum seekers who found themselves shipwrecked on the French Riviera. Initially, these asylum seekers were met with public and political hostility. Despite this initial reception, they managed to convey effective narratives appealing to French sentiments. Presenting themselves as a particular kind of victim, in this instance Kurdish refugees fleeing Iraq, moral sentiments prevailed, and they gained acceptance in the country. In actuality, they were Syrian refugees fleeing the Bashar al-Assad regime, but at this point in time Syria did not appear as evil as the Hussein regime. This short example illustrates "the imaginary of refugees...and hence the idea that they have of the relational and emotional bases on which their requests for asylum will be assessed" (Fassin 2012, 147). It is an intriguing scenario, precisely because it illustrates the interplay and competencies of two spheres of humanitarian actors: the governed and the governors. Humanitarian ideals and governance effectively mask power asymmetries, but, here, humanitarian sentiment is used strategically to support the asylum seekers' goals. Again, this should not be considered a deceptive act but an example of the relationally situated, impression management (Goffman 1959; Berreman 1962). In our contemporary world, refugees and asylum seekers must remain vigilant to the perception wealthy countries hold of various groups to access their basic rights.

Humanitarianism is a system that attempts to regulate people but is also a system that people act within. While the category of victim minimizes capabilities, at times it can also maximize opportunities. Fassin's thorough account of the organizations and governments that set the terms of humanitarian gestures provides a departure point to consider the perspectives of the recipients of those gestures. The interplay between humanitarian spheres requires further analysis; specific in-depth studies are necessary to explore motivations and experiences of those who are governed.

Experiencing Humanitarianism
Life in Camps

Harrell-Bond's 1986 study regarding aid to Ugandan refugees and Malkki's (1995a) work with exiled Hutus could be considered the forerunners of a scholarly analysis of humanitarianism. Harrell-Bond's research took place at a particularly tumultuous time during the initial formation of the refugee camps. The resulting book, *Imposing Aid*, examines the interplay between humanitarian aid providers and recipients. It challenges the simplistic, yet potentially oppressive, understanding of refugees as primarily victims. This understanding, Harrell-Bond (1986, 363) argues, leads to unfettered power imbalances in refugee camps and "lies in the ideology of compassion, the unconscious paternalism, superiority, the monopoly of moral virtue which is built into it." The impact of this underlying logic is a striking power imbalance between refugees and aid workers. Harrell-Bond (2002) argues that, when assistance is provided as charity rather than a way of enabling refugees to enjoy their rights, it is inhumane assistance. This approach forces refugees to ingratiate themselves to their helpers to maintain access to aid and support. Harrell-Bond's work is a powerful critique of a particular moment in the experience of Ugandan refugees: the arrival and formation of the camps. Thus, Harrell-Bond was not able to examine the trajectory the camp experience and the relationship with the label of refugee can take. This, due to the increasingly protracted refugee situations in the contemporary era, merits examination. Refugee camps, as Malkki (1995a) has argued, can emerge as a significant site of social meaning.

Malkki examines the creation of a distinct Hutu social identity in Tanzanian refugee camps. She claims that the camp space fostered the development of a mytho-history that not only gives meaning to exile but re-historicizes the group's exile. Put another way, refugees' mythic history is a political response to the camp experience. This relates to the broader argument put forth in this book: refugees are not necessarily passive in the processes of depoliticization. Despite this impressive theoretical contribution, two aspects of Malkki's work require additional consideration. Fieldwork was performed at a unique juncture in the Tanzanian camps; in 1985, the UNHCR handed control of the camps to the local government (Malkki 1995a). The local district commissioner, shortly after the handover, described the Hutu refugees as economic migrants and challenged the historic events that led to exile. Malkki notes this is a crucial shift, describing the angst participants underwent as their refugee status became precarious. Regardless of this notable change, she approaches it as a peripheral issue in *Purity and Exile* (1995a). Yet several themes in the refugees' narratives suggest a clear effort to claim ongoing international political recognition. For many, the refugee label may be the only form of recognition for past victimization and means to direct international attention to their unmet justice needs. Zetter (1991), for example, argued convincingly that the bureaucratic labelling of refugees has direct implications regarding access to resources and recognition by international bodies. Malkki's work examined the difference between camp- and city-dwelling refugees, but differences within the camps seem nonexistent. It is realistic to assume that different versions of the mytho-history exist, and gatekeepers have a role in regulating which versions were shared. In turn, this may also shed light on why it appears that only males participate in these processes of historical mythmaking. Further, Malkki's (1995a) methodological decision to sculpt data from a variety of sources into composite narratives may have the unintended consequence of collapsing differences within the group (Lubet 2018).

If Harrell-Bond (1986, 1999) and Fassin (2012) emphasize that the status of victim has become a requisite aspect of accessing humanitarian support, Fiddian-Qasmiyeh (2010, 2011, 2014) approaches the idealization of refugees from a slightly different perspective. Her 2014 book, *The Ideal Refugees: Gender, Islam, and the Sahrawi Politics of Survival*, focuses on the performance of Sahrawi female refugees living in Algerian camps as idealized refugee women. Though the book's emphasis is on women, it also provides information regarding men's experiences, and thus offers a slightly broader approach in comparison to Malkki's work. The UNHCR and scholars such as Harrell-Bond (1986) describe the Sahrawi camps as model camps, partly because of the perception that women enjoy high social status. The camps are presented to an international audience as idealized sites of democratic transformation, gender equality, and secularism. The Sahrawi are "good" refugees. Fiddian-Qasmiyeh (2014) illustrates how this image is highly orchestrated—a political strategy that sustains donations to support the camp's livelihood. The camp inhabitants make a concerted effort to present themselves in opposition to the normalized discourse of undeserving Muslim refugees. Undeserving status includes those who do not support Western values: they are undemocratic, uninterested in empowering women, and overly religious. Not content to stop at the merely descriptive, Fiddian-Qasmiyeh attempts to look behind the curtain to understand not only refugees' political manoeuvres but also the consequences. It illuminates that humanitarian support is contingent and conditional, which in turn sustains or creates internal and external power imbalances. Promoting democratic governance either directly or through graduated assistance levels may appear to counter the shift towards expecting refugees to be apolitical. To a degree this is true. However, by celebrating the Sahrawi as exceptional and ideal, the expectation that refugees, and perhaps Muslim refugees in particular, are generally *not* democratic is normalized.

In Australia
While camps are one of the most obvious sites of humanitarian intervention, once refugees are resettled, distinct expectations emerge—albeit still strongly influenced by humanitarian ideals. In Australia, the camp experience, to a greater degree than the events that led to exile, became central to refugees' eligibility for admission. Australia, as with many wealthy countries, strives to be compassionate while still being tough on border security. In Australia, the focus is on accepting the *most* legitimate and *most* deserving, which is almost exclusively linked to extended periods spent in a UNHCR camp. Thus, this policy is used to exclude the many, while also regulating the few who are allowed entry. Humanitarian endeavours are undertaken with the desire to help, however not everyone can be saved. A hierarchy emerges, ranking those considered deserving against those who fall outside the accepted parameters. Once they are resettled, these competing notions create a tension between hostility and hospitality that refugees must negotiate. Thus, the nation and its populace find a "need to redefine the threshold in life

that distinguishes and separates what is inside from what is outside" (Agamben 1998, 131). Becoming morally acceptable to a nation-state requires refugees to transform, but once transformed, their status is still precarious.

Australia is a generous country; 61 per cent of residents donated money to charity in 2020 (Charities Aid Foundation 2021, 11). In relation to all countries analyzed in the 2020 World Giving Index, it ranked fifth and is the highest-ranked "Western" country (Charities Aid Foundation 2021, 7). In Australia, the opportunity to give to those in need is ubiquitous. Outside shopping centres, over the telephone, and via social media, organizations such as the UNHCR, World Vision, Save the Children, Doctors Without Borders, and Oxfam, to name a few, jockey for patrons. Australians are frequently reminded of the suffering in the world and the role they can play in its rectification. Parallel to this culture of giving, an increasingly strict approach to asylum seekers has emerged. While Zetter (2007, 172) has convincingly argued that in wealthy nations there is resistance to both migrants and refugees, this research builds on FitzGerald (2019) and McAdam and Chong (2019) by proposing that, in Australia, this resistance is manifesting itself in a specific fashion. In this context, refugees coming from UNHCR-run camps are elevated above other kinds of migrants and marked out as distinctly more deserving than asylum seekers. The division between refugee and asylum seeker has become reinforced by the recent policies developed by the Department of Home Affairs. Counterintuitively, a strict approach to asylum seekers is framed as a necessary measure to provide adequate humanitarian assistance to "real" refugees. Amid Operation Sovereign Borders, a military-led response to curb asylum seekers arriving by boat, and proliferating offshore detention centres, a specifically victimized refugee becomes the only deserving category. Claiming to want to help those who need it the most affirms a strict admission policy, while reconciling a tightly regulated border with Australia's egalitarian ideals. Watson (2011, 353) describes this paradox of refusing to accept asylum seekers, while supporting the resettlement of refugees from camps, as "the humanitarian defence." This shifts the figure of the refugee from political subject to recipient of charity—refugees become the deserving poor. The power imbalance in which one group is saving/helping another is nurtured, creating "humanitarian fantasies of rescue and salvation that often obscure colonial and post-colonial encounters and global inequality" (Fadlalla 2009, 81). Central to this is the way the deserving or ideal refugee is imagined.

Australia takes a measured approach to refugee integration. Fanjoy et al.'s (2005) work with Australian resettlement representatives suggests that, at the government level, there is an understanding that refugees may need economic support for the rest of their lives. Hutchinson and Dorsett (2012) posit the assumption that refugees experiencing and suffering from ongoing trauma is embedded in Australia's support system. Marlowe (2010) contends that the core of deserving victim status relates to the presumed trauma refugees experience before resettlement. While being traumatized can translate into additional social support through disability payments, and perhaps increased empathy, approaching refugees as traumatized is problematic. Marlowe (2010) found that Sudanese refugees in Australia were met with the assumption they were traumatized. Trauma became an essentialized way of understanding the Sudanese as fundamentally different from Australians. These presumptions functioned as a hindrance to their attempts to integrate in broader Australia; the presumption of trauma marked them as "scarred for life and vulnerable…the refugee master status" (Marlowe 2010, 186). The Sudanese participants in Fanjoy et al.'s (2005) study, while grateful for governmental support, linked a lack of employment opportunities with the widespread perception they were traumatized. These refugees understood their relative exclusion from employment opportunities as contributing to a broader sense of social rejection. This suggests that Australia presents unique challenges to successful integration due to the assumption of indefinite helplessness.

Though trauma has emerged as a benchmark for evaluating claims to social support, it is still subject to suspicion partially because it is not readily observable. This raises an equally problematic aspect: there is not a clear end point to the traumatized victim status. Presuming refugees are traumatized may relegate them to an incessantly peripheral role in Australia. This relates to broader discussion in Australia regarding how cultural difference is understood and negotiated. Hage (1998) argues that, in Australia, the mainstream culture imagines ethnic minorities as largely contained, with little impact on mainstream culture. Further, ethnic groups are expected to express gratitude to the dominant group

(Hage 2003). In a multicultural context, despite aspirations of mutual respect and equality, this allows the dominant group to maintain social power. During resettlement, these two sets of systems, humanitarianism and multiculturalism, begin to mutually re-enforce each other. Both strive to help refugees, but they sometimes have oppressive consequences, relegating refugees to the margins of Australian politics.

Humanitarianism is a discourse of power that governs a sizable portion of the world's population. It is an appealing discourse and feels morally right, but it obscures the reality that these decisions and processes of ranking are largely politically motivated. A critical reading of these multiple situations suggests that humanitarianism results in a global system of governance, where domination and assistance are explicitly linked (Fassin 2012; Harrell-Bond 1986, 1999). Humanitarian reason, in its various guises, appears to be an all-powerful force. Yet this charitable framework has sufficient cracks to allow for action. As Chatterjee (2012, 46) argues, "The everyday reality of subordination, with its attendant ideologies and practices contains elements of submission as well as intransigence." The Bhutanese found ways to move between the contradictory expectations of "refugee-ness" as they strove to become citizens. Between Australia and Nepal, it became clear a Bhutanese refugee community did not simply exist in either site but was created, moulded, critiqued, and reformed by myriad actors who held high aspirations for the Bhutanese. The humanitarian agencies running the camps, the service providers and government staff in Australia, and the Bhutanese themselves attempted to further their respective understanding of what is best for the group. These multiple expectations of the Bhutanese, in turn, necessitated various constructions and performances. The people I worked with were complex and at times contradictory—thus, it was necessary "to represent not only people's most noble characteristics and behaviours, but also the uglier sides...to avoid the tendency in ethnographic writing...to champion our informants' positions" (Faier 2009, 28). Rather than simplifying the participants into the status of "ideal" refugees, I approached them as complex people dealing with a difficult situation. This necessitated a complex, ethically grounded methodology.

Follow the People

Participants had pre-existing ideas regarding the role of the researcher that impacted on my experiences in the field, as well as the information collected. Nearly everyone I conducted in-depth interviews with asked if I was going to be writing a book, or when I would be publishing a paper. I was cautioned by some participants that I needed to present their experience without bias—that I could not be an activist—so my work would be credible. Regarding my perceived role in the community, one informant articulated,

> Without scholars, our children won't believe us. Our experiences will become like a story. We will tell our story to them, and they will think it is just like Slumdog Millionaire, then they will lose their story. Here, they tell me that Mawson Lake [an artificial lake in a housing development, north of Adelaide] used to have water in it, but I can't believe it because I have not seen it. To me it is just "once upon a time" (interview, Adelaide, 2013).

Participants recognized that their stories potentially play multiple roles, not only as an avenue towards future justice but also as a means establishing a particular frame for understanding that will ease their passage into the broader Australian community. As one lady explains, "It is very important that people hear our stories, we did not have a big civil war with lots of death so people will pay attention. We have nothing else, just our stories" (interview, Adelaide, 2013). The sharing of stories, and particularly stories that could serve as evidence for their "genuine" status as refugees, was identified as incredibly important. My research was actively facilitated because enough participants had a personal stake. They saw participation as a chance to tell their stories, promote a political agenda, attempt to increase access to resources, and, for some, contribute to a conversation about the justice needs of Bhutanese refugees. Given the power discrepancy between researcher and participants, voice and representation were central ethnographic considerations. I attempted to integrate the suggestions of participants regarding methodological approaches and direction of research throughout my fieldwork. Although it is crucial to include the voices of participants, they represent only one piece of a complex puzzle (Lareau 1996). Many of the participants actively sought to navigate my research in a direction that supported their personal agendas—as could be expected with any large, diverse group negotiating constraining frameworks of social participation. Study participants are likely to want certain issues to be developed, reflecting their views and the way they see themselves in relation to the world (Berreman 1962). Recognizing that participants are active agents in the manner they present themselves is crucial. Thus, to do justice to them, I needed to go beyond simply reporting their views.

"Go to Nepal, then you will understand" was the common refrain from participants when I began my fieldwork in Australia. Over eighteen months, this research moved from Adelaide, South Australia, to the refugee camps in eastern Nepal, before circling back to my first site. This circle brought into sharp focus the fact that, though geographically distant, the group was exceptionally interconnected. Performing research both in the camps and with the resettled Bhutanese refugees was a strategy designed to "follow the people," an approach championed by Marcus (1995). A multi-sited approach facilitates the contextualization of experiences and understandings across the multitude of spaces societies inhabit (Marcus 1995). It is a specific strategy for designing ethnographic research involving "strategies of quite literally following connections, associations, putative relationships," and putting those links at the core of the ethnographic inquiry (Marcus 1995, 97). This approach proves particularly fruitful for studying the experience of migrants (Leonard 2009) and refugees (Colson 2007), because it is effective at "investigating precarity under neoliberalism as the method mirrors the social realities" that families living in different countries experience (Francisco-Menchavez 2020, 56). Further, as Kenny and Lockwood-Kenny (2011) observe, pre- and post-resettlement accounts are relatively rare in the existing refugee studies literature. While a multi-sited orientation is not necessarily more holistic than a single site, it helps to orientate the research very quickly towards the diverse scale of structures that impacts on everyday life. While focusing on these specific, local behaviours, this approach also demands the recognition of broader social, cultural, and political locations (Gupta and Ferguson 1992). Refugee camps, far from existing apart from the world, are firmly embedded in a globally linked humanitarian system of governance.

Multi-sited ethnography has been criticized for potentially not providing enough depth due to potentially abbreviated time periods; "this type of research implies moving around and 'following' horizontally, there is little time for staying put and following vertically" (Falzon 2009, 7). In the Bhutanese context, though the fieldwork sites were physically distant, they were intrinsically linked. Before leaving for Nepal, I spent an afternoon with some of the Bhutanese, writing postcards bearing images of Adelaide and taking photographs I was tasked with delivering. Once in the camps, I would track down a family member and give them the photo or message. Immediately, the recipient would use their cell phone to call the sender, laughing that the photo or postcard had arrived. It made for a warm reception while also shattering any assumptions the Bhutanese refugee camps were isolated, existing a world apart. There was a rapid exchange of information and ongoing connections that the camps had in relation to the wider world. Bearing photographs and postcards from friends and family resettled in Australia helped establish a certain rapport that may have taken longer to develop otherwise. It facilitated my research in unanticipated ways, providing different results than a single-sited study. It illustrates not only the different geographic spaces the Bhutanese experienced but also the way ideas, norms, and values become contingent on the setting. For the Bhutanese, it proved practically effective because their experiences have been multi-sited. Everyone resettled in Australia has spent time in Nepal—though not all necessarily lived in the camps. In a sense, the multiple sites articulated I was working with the same population, while it simultaneously revealed I was not working with one coherent community.

In addition to the expectations of international and state-level institutions, the role of Bhutanese leaders is central in understanding their experiences. Any research with refugees must be done respectfully, with an awareness that "dignity" can be precarious during displacement (Banki and Phillips 2017). Initially, I attempted to conduct collaborative, participant-led research—an iterative approach that ideally would help lessen the power imbalance between researcher and participants. However, "researchers' good intentions toward community participation in research practice may bump against limits that are the result of power hierarchies in refugee communities" (de Smet et al. 2021, 2). The refugee leaders—themselves from diverse backgrounds and sometimes with competing goals—had clearly developed political agendas and worked tirelessly to promote them. In doing so, they sought to construct a particular image of the experience of exile and what it meant to be a Bhutanese refugee. Some aspects of this image reflected international values, particularly when these buttressed existing social structures. Others, such as the caste system and particular marriage customs, did not reflect such values and seemed to threaten the Bhutanese leaders' ability to maintain an image of "model" refugees.

Early on, service providers flagged issues that would have been masked for several months. The caste system was one such matter. Service providers inadvertently aided in revealing the ways the Bhutanese managed their identities, which iteratively became the focus of this research. Initially, I was reluctant to analyze the caste system—worried it would throw a poor light on the Bhutanese. Yet failure to analyze it would have meant prioritizing moral evaluations that regarded the caste system as a "problem" over the importance the system held for the group. It would have contributed to the too-frequent approach to refugees that assumes a lack of political institutions: that displacement has stripped refugees of everything that makes them complex, complete humans. It was tempting to

> *fall into deep sympathy with the people we are studying, so that while the rest of the society views them as unfit in one or another respect for the deference ordinarily accorded a fellow citizen, we believe that they are at least as good as anyone else, more sinned against than sinning. Because of this, we do not give a balanced picture (Becker 1967, 240).*

During my time in the field, it became evident that studying moral systems demanded the utmost attention because nations, international organizations, and local service providers subjected the Bhutanese to such a degree of evaluation. Similarly, the Bhutanese had their own strong systems of moral governance. To understand the Bhutanese, the larger environment that regulated, governed, excluded, and integrated them, as well as their own social hierarchies, required analysis. Despite a researcher's best intentions and efforts, research can be disempowering for some participants (Lokot 2019), but it can also be empowering:

allowing refugees to negotiate categories of belonging, to push against stereotypes, and frame their relationship with the nation-state.

In addition to participant observation, I conducted interviews and focus groups. Interviews generally lasted two hours, though some stretched over multiple days. I ended interviews with an open-ended question (Spradley 1979; Leech 2002): "Is there anything you would like to add?" The response was always the same: "I want to convey my thanks to the UNHCR [or Australia]." Though the Bhutanese strove to present themselves as equals to the international organizations that managed the camps, and to the broader population of Australia, they found themselves in a paradoxical relationship. While it was reiterated that all humans are equal, their status as recipients of compassion demanded humility—positioning them as slightly unequal to those with the power to give (Fassin 2012).

Focus groups arose organically among participants and were not preselected. I would often plan on speaking to a specific person only to arrive and find two, three, or four other people who also wanted to talk. These groups made for heady discussions as ideas and interpretations were debated among participants. The groups often also brought social norms into sharp focus. During a family interview (Adelaide, 2012), I asked the youngest daughter what she thought about the refugee camps she had grown up in. "The camps, they are just horrible. They are—" Her father interrupted, "The older people have knowledge the younger people can't explain, our real problems, the census, and our government's policy." This interaction articulated the intergenerational tensions that participants felt resettlement accentuated. It also illustrated an assumption I consistently came up against: senior males would be the "natural" choice as participants. This was an assumption held by both men and women, regardless of age or caste, with very few exceptions. When I visited people's homes, women served tea while men answered my questions. I was viewed as someone who moved in the public, political realm of men, while women—with a few exceptions—circulated in the private, domestic arena.

In Australia and in Nepal, this was overcome by my pregnancy. Women, who previously were reticent or deferred my questions to men, began speaking to me. At first, they provided pregnancy advice regarding the specific fruits and vegetables I should consume or avoid.

In the camps, I was quite ill, and women extended considerable kindness due to my condition. They shared folk remedies and allowed me into their homes to rest while recovering from nausea. Soon, they revealed considerable details about their lives that often began with their experiences of motherhood. It was through these conversations I learned about their experiences as refugees and their hopes for the future. In Australia, the women similarly doted over me, preparing soups to reduce indigestion or providing small gifts of fruit to help the growth of my baby. They eventually threw a baby shower to celebrate the birth of my daughter. Thus, this unexpected life event provided a "bridge to humanity": a means of creating a personal relationship with women (Grindal and Salamone 2006). Similar to the experience of Basnet et al. (2020) with the Bhutanese in New Zealand, this unexpected event opened lines of ethnographic inquiry that otherwise may have been difficult to access.

Lubet (2018) argues that ethnographers must rely on a variety of "fact checking" processes to ensure their conclusions are robust. While participant observation was the primary approach, this research was enhanced by several other qualitative techniques and was supplemented with multiple quantitative surveys (see Chapter 3). Pulling together these diverse, yet complementary, methodological tools, I hoped to increase the academic rigour of emerging interpretations. Mixing approaches can yield more diverse viewpoints (Flick 2004; Denzin 2012) and support inferences drawn from observation (Lubet 2018). By combining qualitative and quantitative tools, this research aimed to triangulate my data to "obtain a better, more substantive picture of reality, richer more complete array of symbols and theoretical concepts and a means of verifying many of these elements" (Berg 2004, 4).

The role of the researcher in relation to the expectation of power raised ethical concerns during fieldwork. The expectation of power was most pronounced in relation to institutions that were perceived as being "Western." This is not unique to the Bhutanese camps. MacKenzie et al. (2007, 303) suggest, "Some participants may have unrealistic expectations of the benefits of the research, believing that researchers have the power to influence legal or resettlement processes." Several participants sought me out, hoping I could facilitate or expedite their resettlement process. I reiterated I was not associated with the UNHCR or the IOM, and that our interactions had no bearing on the outcome of their resettlement process. What these interactions did illustrate was just how powerful the humanitarian actors in the camp setting were—the fate of thousands hinged on the decisions they were making.

Repaying a group for hosting a researcher for an extended period of time is perhaps an insurmountable task. People spent hours of their lives answering my questions, explaining the obvious, and teaching me all manner of things. They fed me, provided me with an incredible volume of chai, welcomed me into their homes, invited me to social and religious events, and, together, we shared many of life's most precious milestones. Reciprocity is imperative for social research to maintain its ethical foundation. I have fulfilled the hope of many that their experiences would be shared with others through academic and practitioner-focused publications, as well as at conferences. During fieldwork, I was a volunteer for the City of Salisbury and facilitated a craft group for elderly Bhutanese. Though the City of Salisbury donated some supplies, I supplemented the haberdashery with specific items participants requested and items I thought they might enjoy. I also assisted in English lessons for adults, as well as the cleaning of the community centre. I frequently found myself explaining bureaucratic processes, helping with minor translation during training courses, and arranging service providers to speak with the group. For example, the group raised the issue of wills and probate law. Some misinformation had been moving through the group regarding burial requirements and the validity of wills in Australia. Many were concerned that cremation was not allowed in Australia. I arranged for the Legal Services Commission of South Australia to do a month-long series of talks to provide the group with accurate information. There were smaller acts of reciprocity as well, such as acting as a referee for informants that were having difficulty procuring housing due to a lack of local contacts, and providing character references on employment applications.

This book argues that, as refugees interact with humanitarian actors and move within humanitarian policy frameworks, they must adapt and "learn" to be a particular kind of refugee. The experiences of the Bhutanese illustrate the complex strands of power that intertwine to limit the scope of people who deserve compassion. In this context, refugees' ability to make claims based on the right to sanctuary becomes more precarious. While the Bhutanese have found ways to work creatively in these constraining frameworks, their experiences illustrate that the well-meaning discourse of humanitarianism soothes the conscience of the global powers as they face ever-increasing evidence of the injustices that nation building causes and national boundaries sustain. More and more, the role of the few refugees who are granted the opportunity to resettle is to resolve the moral dilemma that the daily reminders of global inequalities present.

This chapter has outlined the ways that humanitarian assistance and the expectations it entails of refugees have evolved. The next chapter of this book introduces the movement of the refugees from Bhutan, through India, and into Nepal as a means of situating the political context that led to exile. While this is linked to colonialism and global processes, the chapter underscores that nation-states in the Global South also exercise a degree of agency in the management of refugee flows. The next section of the book examines the experience of "learning to be a refugee" (as a participant astutely described it) in refugee camps. This learning process is analyzed across two chapters: Chapter 3 deals primarily with the performance of the "ideal" refugee, as participants attempt to reflect norms, values, and behaviours of international benefactors. The subsequent chapter analyzes the strategies participants employ to mask behaviours deemed by humanitarian representatives as unacceptable. Through an examination of how institutional expectations are being performed, contested, and negotiated, the tension between structures of humanitarian governance and individuals in the camp settings can be analyzed.

The final section of the book focuses on Bhutanese resettled in Australia. Though strong similarities exist between the ideology of the UNHCR and the government of Australia (namely equality and democratic governance), there are pronounced differences. Chapter 5 introduces the political and public environment of Australia between 2012 and 2014. In Australia, the ability to manage those deserving of compassion can have a powerful legitimizing effect. Chapter 6 examines the interplay between humanitarian ideals and domestic politics, arguing that governing refugees can be analyzed as a form of domestic humanitarianism. The Bhutanese conform to and push against humanitarian ideals: Chapter 7 illuminates some of the consequences of the humanitarian discourse and the constricted framework towards participation it allows. The concluding chapter argues that, though the Bhutanese are a particular group, as refugees in a globally connected world, their experiences illustrate the ways that humanitarian gestures ultimately reinforce social inequality. As a system of governance, humanitarianism transforms refugees into humanitarian subjects that are distinct from citizens. It is through this process that international organizations and national governments gain credibility, while absolving themselves of the inequalities deeply embedded in the nation-based system. Always supplicants in a global reconstruction, refugees have become manageable, yet in a way that continues to restrict their freedom.

2
On the Fringe of Empires

TO UNDERSTAND how individuals and groups respond to the global discourse of humanitarianism, we must understand the context of their own history and culture. Bhutan is often referred to as a hermit kingdom, yet the culture that exists today is "not timeless and pristine objects [but] products of restless operations of both internal dynamics (mostly local power relations) and external forces (such as capitalism and colonialism) over time" (Ortner 2006, 9). Thus, as isolated as the nation of Bhutan and the people within its boundaries may seem, it is possible to trace broader connections: a globally linked history. Wolf and Eriksen (2010, 3) show convincingly "humankind constitutes a totality of interconnected processes...this holds true not only of the present but also of the past." The implications of these links, though noteworthy in their own right, take on additional significance in the context of the exiled Bhutanese.

Malkki (1995b) argues that refugees are frequently conceptualized as being without history, an understanding that naturalizes their peripheral status. To overcome the lack of history, it is crucial to historicize the processes leading to the formation of refugee populations (Malkki 1996). After leaving Bhutan, the refugees found themselves as pawns in regional power plays to affirm state sovereignty. Additionally, each of the nation-states that the Bhutanese refugees moved through considered their own diplomatic concerns and possibilities, exerting agency even when faced with few options. Adamson and Tsourapas

(2020) demonstrate a need to carefully analyze approaches to migration in the Global South to help balance a literature that tends towards the experiences of the Global North—developing a more comprehensive understanding of migration management policies. For the Bhutanese refugees, this had the effect of further positioning them outside the nation-state while simultaneously making them legitimate recipients of humanitarian assistance. The people that fled Bhutan are not an accident of history but rather the product of a globally linked social system that both generates and perpetuates refugee populations.

This chapter introduces the geographic, political, and social context of Bhutan in relation to the larger global processes of colonialism, postcolonialism, and regional power dynamics. Analyzing these interrelated processes allows us to see the refugees' historical construction that, in turn, positions them as targets of humanitarianism. In other words, it becomes imperative that humanitarian intervention's relationship to the historical legacy of colonialism and the nation-state is analyzed. This chapter demonstrates the ways refugees provide an effective mirror to examine humanitarian responses as a global project interacting within distinctly regional contexts.

Bhutan and the British

When I tell people I work with Bhutanese refugees, I am often met with exuberant statements about Bhutan. In Australia, it was clear that, though few visited the country, Bhutan holds a place of esteem in the minds of many people. It is popularly imagined as the last Shangri-La, a mountainous kingdom, free of the corrupting influence of the West, inhabited by a people more concerned with happiness than material consumption. For those that follow events a bit more closely, it has been hailed as a Buddhist kingdom that saw a peaceful transition to democracy in 2007. The most frequent word Australians used to describe Bhutan was "isolated." While the British never directly colonized Bhutan, the group that became refugees is part of a larger colonial legacy that reflects ideas regarding national borders and the strategic utilization of ethnic groups to consolidate power. During Bhutan's transition from a multi-ethnic, absolute monarchy to a more homogenous democracy, a narrow vision of national identity was viewed as necessary to maintain sovereignty.

In order to begin to understand how refugee populations are generated, one must examine the peculiar coordination of space, time, and people (territory, history, and society) (Foster 1991). Geographically, before 1865, Bhutan's territory stretched into northern regions of the British Raj. The term *territory*, rather than state, is used deliberately, for "on land...there was no precise line of demarcation but rather a band, zone, or interval...a ruler ruled as far as he could collect taxes and maintain order" (Lewis 2002, 127). At this point, Bhutan's southern border extended an additional 4,400 square miles across states already under British control. This relatively flat, malarial-prone land provided seasonal grazing for Bhutanese herders, was a source of tribute, supported trading posts, and supplied slaves to Bhutan (Phuntsho 2013). These benefits were invaluable to the Bhutanese due to the otherwise mountainous landscape. The British, on the other hand, viewed this land as underutilized and problematic due to the lack of a precisely delineated border (Rennie 1866). As the neighbouring states in British India began to actively utilize similar zones for lucrative tea cultivation (Besky 2013), this area became desirable. The previously tolerable relationships between Bhutan and the British began to break down as frontier incursions by both groups increased. The stated motivation for acquiring these tracts of land was to protect the inhabitants of northern British India from Bhutanese attack. However, this claim is dubious and was likely opportunistic, based on disputes averaging roughly one case a year (Gupta 1975). It seems more likely this land was pursued due to its potential for tea cultivation (Sarkar and Ray 2007). Regardless, these border disputes provided the justification for a full-scale attack that culminated in the 1865 Duar (or Anglo–Bhutanese) War. Close to a fifth of Bhutan's most economically viable land was ceded to the British. Suddenly, the centrality of national borders to laying claim to resources became not only normalized but an urgent necessity. Regions, if not defined and delineated in a manner that mirrored the British model, could be lost.

The end of the Duar War left Bhutan embroiled in a civil war. Infighting between local governors led the British to describe the country as "an incomprehensible hierarchy" lacking a clear leader (Ansari 2012, 51). In 1903, the British governor in neighbouring Darjeeling proposed a mission to Tibet to counter the perceived threat of Russian imperial expansion (Aris 1994). Ugyen Wangchuck, a powerful Drukpa governor in central Bhutan, believed it would be in Bhutan's best interest to assist in this mission. After the mission, Wangchuck was invited to Calcutta (now Kolkata) and treated ceremonially as Bhutan's ruler. The British Empire conveyed its formal appreciation for his diplomatic work by presenting him with the title, "Knight Commander of the Indian Empire" (Aris 2005, 90). Though the crowning of Wangchuck in 1907 as the first king of Bhutan was not done at the behest of the British, their support was instrumental. This reinforced a hierarchy in Bhutan, which positioned the Drukpa as the political elites. Though there are multiple ethnic groups in Bhutan, Drukpa cultural history and social norms emerged as the hallmarks of a national identity. The contemporary, popular imagining of "who" constitutes the Bhutanese reflects these historic events—Buddhists wearing the *gho* or *kira*, speaking Dzongkha amid whitewashed temples.

As the government was consolidating, the redevelopment of Bhutan's economic system was simultaneously taking place. Previously, serfs belonging to a multitude of ethnic and Indigenous groups worked the lands and gave tribute in the form of goods to the landowner, who passed a portion to the government. Due to taxes being paid in kind, "the central government had few sources of revenue" (Hutt 2005, 80). To develop a tax system based on money, rather than goods, the remaining southern lands needed exploitation. Across the Himalayan foothills, the British imported nonlocal labour—a group classed as Nepalese labourers—to cultivate tea and other cash crops (Besky 2013). The Nepalese represented a myriad of ethnic backgrounds and several languages, however they tended to be Hindu, and Nepali was frequently the lingua franca. The British imagined the Nepalese as being settled agriculturalists, "industrious, loyal, and easy to control," and rewarded them with a higher wage than local workers (Besky 2013, 80). While some migration into southern Bhutan was occurring organically, much of it was assisted by Drukpa governors to increase the kingdom's capital and their own relative power.

The group of Nepalese ancestry that settled in Bhutan became known as Lhotshampa,[1] literally "people of the southern border." Agriculture in the south proved to be a valuable source of government income, as it was the area where cardamom, oranges, and other cash crops could be grown. These economic contributions were welcomed and encouraged by powerful elites. As Hutt (2003, 80) observes, "It is, therefore, very clear that the main motives for encouraging Nepali settlement in the south were economic ones." Not everyone benefitted from changing land use patterns. A survey conducted by White in 1905 suggests that the lifestyle of the nomadic Bhutanese from the northern districts was undergoing considerable change due to interactions with Nepalese settlers (White 1909). The sedentary lifestyle that intensive agriculture demanded fundamentally altered the way land was being used in the south. No longer could Bhutanese herders move their livestock from the northern foothills to winter on the more temperate plains. This further stratified the groups within Bhutan into ruling elites, nomadic pastoralists, and immigrants participating in capitalist endeavours. The British reinforced a political hierarchy that placed the Drukpa as rulers and the Nepalese as a migrant, labour caste different from the local population. It also began to solidify the Nepalese as a unitary group, despite the many tribal, caste, and religious differences within this broad label.

These stratifications have strong parallels with other colonial situations. In Malaysia, the British utilized Chinese and Indian labourers while accommodating political elites (Ong 1987). A similar process occurred on the subcontinent. The British in Darjeeling actively encouraged Nepalese migration, as they had a vested interest in seeing a protectorate population settle the borderland between the Empire and spaces of perceived Sino-Tibetan influence (Ghosh 2010). Colonial rulers "sought to create a subservient Nepali land-owning class in order

---

1. There is no single term that all the participants accepted regarding how they wanted to be identified. The term "Lhotshampa" was viewed by some as representing an attempt by Bhutan's government to minimize their Nepalese ancestry, while others felt the term "Nepali-speaking Bhutanese" placed too much emphasis on the Nepalese aspects of their identity. Due to a lack of consensus, I use the terms Bhutanese and Lhotshampa interchangeably, particularly when discussing their experience in Bhutan and the refugee camps. For the discussion of their experiences in Australia, I use the term Bhutanese, as this is generally how they refer to themselves when interacting with service providers and the broader Australian population. Nelson and Stam (2021) argue the refugees' experience of displacement has necessitated a degree of flexibility in identification.

to counteract the traditional predominance of the Tibetan and the Bhutia landed aristocracy" (Dasgupta 1999, 55). This facilitated colonial management. The local population and the immigrant population were used in different ways, accentuating perceived differences, while the ruling elites were given some degree of licence. This practice is very much akin to the formation of caste-type groups by Belgium in Burundi: a stratified society became a means of extracting a maximum amount of taxation while maintaining comparable control (Malkki 1995a). Similar to the Tutsi and Hutus in Burundi, differences in Bhutan would become reinforced and exploited. These "colonial efforts to fix racial boundaries and spatially enclose groups" (Bissell 2007, 185) were to have future consequences and an impressive ability to reincarnate themselves, as the following chapters will illustrate.

After the coronation of the first king in 1907, the country faced major challenges in terms of transitioning the economy away from its earlier feudal structure, fostering a sense of national identity, and maintaining independence in a strategically significant region (Karan and Jenkins 1963). Yet, even facing these changes, Bhutan enjoyed a relatively stable period. This period came to an end in 1950, when China annexed Tibet. To the north, the massive sovereignty battles that played out in Tibet were prompted by an external power: China (Gallenkamp 2011). To the west, a similar battle played out in the Buddhist kingdom of Sikkim, though this time the catalyst came from within. Similar to the ruling lineage of Bhutan, Sikkim's ancestors came from Tibet in the sixteenth century and founded a monarchy stemming from Buddhist ideals. Sikkim enjoyed a close relationship with Bhutan, bound not only by a shared religious past but intermarriage of royal families. The current king of Bhutan has a Sikkim princess as a grandmother. Over the span of a hundred years, the Nepalese population in Sikkim, encouraged by the British to work on tea plantations, became the most dynamic aspect of the economy and the demographic majority (Rose 1963). The Sikkim royal family came to represent the minority of the population and reacted by taxing the Nepalese at higher rates and limiting representation in government, increasing friction between the two groups (Levi 1959). In 1975, a strong push by the Nepalese majority led to the referendum that abolished the monarchy and merged the country with India (Gupta 1975). Another key political event was the violent

Gorkhaland independence movements in Darjeeling between 1986 and 1988 (Thinley 1994; Evans 2010a). This movement, led by ethnic Nepalis, was an attempt to expand the boundaries of present-day Nepal to include contemporary Indian states. It "must have played a major part in convincing the Bhutanese government that political activity among the Lhotshampa should be prevented at any cost" (Hutt 2003, 195-196). Bhutanese government officials witnessed multiple countries in the region undergo shifts that stripped virtually all power from the traditional political elites. They also saw these transitions seemed to be spurred by groups of people that spent generations in the respective country but were perceived to be "immigrants" who had not properly assimilated.

Building Nations
Though taxes from the southern foothills flowed north, the movement of people from the south to north Bhutan was strictly regulated. An internal boundary that restricted people of Nepalese ancestry to the southern regions remained in place until the 1970s (Hutt 2003; Thinley 1994). There were government efforts as early as the 1950s to homogenize the culture of Bhutan by converting the Lhotshampa to Buddhism (Royal Government of Bhutan 1953). By the 1980s, the large, predominantly Hindu population occupying the southern border was perceived as a threat to the emerging Buddhist kingdom. This was partially because the south Bhutanese represented a blurring of cultures rather than the firm, geopolitical boundary Bhutan was attempting to establish. The lessons from the Duar War illustrated the dangers of having an undefined border. Bhutan responded to these perceived threats by encouraging a very particular image of "who" was Bhutanese, undertaking a process of nation building based on the principle of "one nation, one people" (Royal Government of Bhutan 1993). These policies favoured the Drukpa majority, while initially downgrading, then alienating, the rights of minorities (Gilroy 1990; Hutt 2005; Whitecross 2010). This ethnically exclusive nationalist policy strove to create a homogeneous Buddhist kingdom that dressed, spoke, and worshipped alike. Those that fell outside the national image were considered at best to be deviant and, at worst, terrorists (Royal Government of Bhutan 1992).

The Bhutanese of Nepali ancestry had been in the country for several generations, but they remained a distinct ethnic group. Further, they effectively controlled the majority of Bhutan's agricultural land, an invaluable resource in a mountainous region (Hutt 1996). These attributes began to be seen as a threat to Drukpa elites, and the government responded by encouraging a national identity that mirrored the ethnic stereotyping and hierarchies created during the colonial period. Several related events, along with more integration efforts, would prove to be catalysts for exodus: the Marriage Act of Bhutan, 1980; the Bhutan Citizenship Act, 1985; multiple pro-democracy pamphlets published between 1988 and 1989; and the 1988 Census.

For Hindus, marriage ideally is arranged based on caste hierarchies. The new bride must be of the same caste and ideally reside in a different village. After marriage, she will live with the husband's family, reflecting a strongly patrilocal social structure. In Bhutan, the different village was not necessarily inside national boundaries. Wives could either come from Nepal or be Indian nationals of Nepalese ancestry. As a result of the Marriage Act of 1980, a citizen married to a noncitizen lost their eligibility for promotions and access to loans or goods such as livestock, and the educational opportunities of their offspring were curtailed. These provisions were viewed as having a disproportionate impact on the south Bhutanese. This was likely the government's intention.

Bhutan's National Assembly adopted the Citizenship Act in 1985, and a census based on this act that sought to identify Bhutanese nationals was implemented in 1988 (Hutt 2005). Previously, anyone whose father was a Bhutanese citizen was able to become a Bhutanese national. After the Citizenship Act, only those born to parents who both held Bhutanese citizenship would be citizens by birth. These changes were perceived as targeting the ethnic Nepalese, whose marriage practices tended to occur along caste lines that frequently crossed national boundaries. The Citizenship Act was particularly problematic in that it was retroactive in scope: only those who could produce tax receipts dating specifically to 1958 could be considered citizens (Lee 1998; Saul 2000). Other documentation, including land titles, marriage certificates, government-issued passports, or tax receipts from other years were considered insufficient. The 1958 Nationality Law did not have specific provisions regarding language abilities or knowledge of specific cultural mores. The Citizenship Act further enforced Drukpa cultural norms through the policy of national etiquette called Driglam Namzha (Hutt 2003). The etiquette is extensive, including dictating which clothes are appropriate, what language is allowed in public, and how superiors are to be addressed. The Citizenship Act stipulated that people must be able to speak, read, and write the Drukpa's language, as well as have a good knowledge of the culture, customs, traditions, and history of Bhutan. These were difficult requirements to meet for most people along the southern foothills, due to an education system that was based on the Nepali language.

Wearing particular clothes, having to observe Drukpa social norms, and speaking a different language were considered unbearable: the south Bhutanese protested, demanding democratic reforms. Several associations founded by south Bhutanese, including human rights groups and student organizations, staged

mass public demonstrations in southern Bhutan in September and October 1990 that were unprecedented in the kingdom's history. The demonstrators submitted a list of demands that was clearly influenced by the wave of democracy movements and human rights activism that had swept across eastern Europe, and had very recently reinstated a multi-party democracy in Nepal (Hutt 1996, 406).

"Seditious pamphlets" (as described by the Royal Government of Bhutan in 1992 and 1993) appeared, and eventually forty-five people were arrested for what one arrestee described as "raising our voice for democracy" (Gazmere male, age 40s, interview, Adelaide, 2012). Arbitrary arrests increased until thousands of south Bhutanese were imprisoned (Amnesty International 1992a). The arrest of dissenters (and reported deaths while in custody) reverberated through both the north and the south. The involvement of established, high-ranking public servants and university professors was interpreted by the government as demonstrating that south Bhutanese could not be trusted—even those in high positions were not loyal citizens. The south Bhutanese interpreted the morphing policies as an attempt to force out those who spoke against the government. The south Bhutanese suspected the government was attempting to make them submit to the Drukpa culture or leave the country (Hutt 1994, 11). Tension between the two parties continued to mount, becoming more and more violent. Eventually, an arrestee's wife contacted Amnesty International to intervene on her husband's behalf, hoping to secure his release from prison. Amnesty International performed an investigation, finding his imprisonment to be largely arbitrary.

> The individual violent crimes...were all committed in April–June 1990, six or more months after they were detained. It also examined the booklet Bhutan: We Want Justice *and concluded that it did not contain threats of armed uprising against the state or the advocacy of violence (Amnesty International 1992a, 9).*

The efforts of Amnesty International led to the release of 313 political prisoners who had been shackled in solitary confinement for criticizing integration policies (Amnesty International 1992b). When the political prisoners were eventually freed through the efforts of Amnesty International (1,500 were pardoned by the king in 1992 after Amnesty International's visit), many found their property confiscated and their citizenship revoked. Saul (2000, 347) argues, "The mass exodus of south Bhutanese was immediately caused by discriminatory nationality laws, overly zealous cultural protection laws and laws repressing democratic dissent." The US Department of State (1990) similarly maintains that the Royal Army and Royal Police of Bhutan committed human rights abuses such as arbitrary arrest and rape. The refugees view the government's policy of integration as a concerted effort to ethnically cleanse part of Bhutan. The government of Bhutan maintains that illegal immigrants who squatted in southern Bhutan left willingly during this period. The refugees maintain that systematic harassment, including rapes, physical attacks, arbitrary arrests, and the kidnapping of family members forced them to flee. The US Department of State (O'Brien 2010), Human Rights Watch (2003), and Amnesty International (1992a, 1992b) support the refugees' version of events.

State Power Plays

Where refugees end up after their flight is the consequence of complex political and historical factors. Countries consider their own sovereignty, as well as the financial expense they may incur or the support they may receive. Particularly when there are asymmetries between states, refugees can be used to further both regional and international political agendas through "migration diplomacy" (Adamson and Tsourapas 2019). Increasingly, accepting refugees can provide emerging governments with a chance to gain credibility by mirroring the contemporary emphasis on humanitarian values, coupled with the concerns about irregular movement across borders that are prevalent in Western politics. It also can curry favour with wealthy Western countries that want to ensure refugee flows do not end up on their doorsteps, so to speak. While wealthy countries gain moral credibility by accepting a few refugees for resettlement, poorer countries that host refugees can leverage their position to increase their economic and/or political position (Adamson and Tsourapas 2019). Refugees are frequently described as excess, undesirable populations (Agier 2010), but as this section illustrates, the Bhutanese refugees were utilized as pawns in complex power plays between states. Nepal was not merely a passive recipient of a flow of Bhutanese refugees but actively negotiated with multiple state and international actors to assert its sovereignty.

For the refugees fleeing Bhutan, India was the first country where they sought sanctuary. The border between Bhutan and India is governed by the 1949 Treaty of Friendship and was formally delineated between 1974 and 1983. Bhutanese citizens are not required to have a passport or visa to enter India. Further, India is not a signatory to the 1951 Refugee Convention. Thus, India's 1946 Foreigners Act and 1955 Citizenship Act govern noncitizens entering the country. These acts stipulate that entering India without the proper visas makes such trespassers liable to deportation. The Bhutanese camping on the India border in the late 1980s and early 1990s technically were protected under the Treaty of Friendship. However, they were still vulnerable to deportation and forced eviction. Estimates vary regarding how many refugees remained in India. Human Rights Watch estimates between fifteen thousand to thirty thousand, but it is not clear if both Lhotshampa and Sharchopas are included in this total (Ridderbos 2007). Compounding on this vulnerability were lingering effects of the Gorkhaland movement that saw "about 200 people" die in North India as members of the Nepali diaspora violently attempted to expand Nepal's boundaries (Hutt 1997, 130).

During this period, India was countering multiple separatist movements across the country, including in Assam and West Bengal—both of which border Bhutan. The United Liberation Front of Assam formed in 1976 and has since then undertaken an armed struggle to gain independence from India. India claims the organization is a front for the Pakistani intelligence agency and considers it a terrorist group (Prakash 2008). An additional group, the Bodoland Movement, claims to represent the Bodo ethnic group historically indigenous to the foothills of northeast India, Bhutan, and Nepal (Mullick 2001). It also seeks autonomy from India through armed struggle. Both of these groups established bases along the Bhutan–Indian border in the 1980s and 1990s, moving between the countries to avoid detection. These bases functioned as staging grounds for attacks against Indian police, government officials, and infrastructure. The groups also targeted various ethnic groups—including people with Nepali ancestry—originally brought to the area by the British to work tea plantations (Prakash 2008). Both groups are antagonistic to those described as immigrants in the region and attempt to legitimize a claim to the political and geographical space through ethnic longevity. There has been speculation by the refugees that these groups, due to their strong anti-immigration stance, worked with the Bhutanese government to force the south Bhutanese of Nepali ancestry out of the country (interviews, Nepal and Australia, 2012–2013). Again, the violent acts by the Gorkha National Liberation Front in the region may have also made a reception in India unlikely.

In addition to the groups based in Assam, West Bengal housed several Maoist groups that advocated armed struggle to overturn the Indian government. In the early 1990s, the Indian military staged operations against the United Liberation Front, the Bodoland Movement, and the Maoists by dismantling all camps along the border. The government did not treat the camps housing Bhutanese refugees (that were not advocating armed struggle against India) any differently than the camps that housed anti-Indian militant groups. By forcibly removing these groups and the refugees, India asserted its sovereignty over a precarious border region. Indian security further exerted its power over establishing who was a legitimate citizen of Bhutan or India by forcing the exiled Bhutanese to Nepal. As the following section will demonstrate, this was a strategy to highlight India's power in the region.

India seemed to view the Bhutanese refugees as yet another group threatening state power in an already fraught border. Due to their Nepali ethnic heritage, the refugees did appear to have a potential role to play in political relations with Nepal. Though the 1950 Treaty of Peace and Friendship between the Government of India and the Government of Nepal stipulates the free movement of people and goods across the border, by the late 1980s this arrangement was coming under considerable strain. As an increasing number of Hindi-speaking Indians settled within Nepal's borders and sought Nepalese citizenship, citizenship became more politically charged. Nepal introduced a transit visa system for Indian workers, attempting to supplant the existing model that allowed for free movement. India responded by supporting the expulsion of ethnic Nepalese from Indian states. These ethnic Nepalese, aside from being viewed as equal to Indian nationals under the treaty, often had been born in India (again, having been brought to the region by the British as labourers). India's governmental support for their expulsion illustrated the manipulation of ethnic or national categories for political gain (Nath 2005).

From March 1989 until June 1990, India imposed strict economic sanctions, unilaterally closing its borders with Nepal. Tensions between the governments escalated. Refugees shuttled to Nepal by Indians during this time had few options: Indian border police hindered their re-entry. Similarly, the Nepali government had few options, as India crippled Nepal economically (Hagerty 1991). The Nepali government was effectively forced to accept political arrangements dictated by India (Hagerty 1991). Maintaining that the people India delivered into Nepal did not have a claim to be in the country may have been a small effort by the Nepali government to wrestle back political power over its borders. By the end of 1990, several hundred refugees were in Nepal. The Bhutanese refugees became pawns in a struggle not only over land in contested border zones but in larger political arrangements between countries.

The people who initially fled Bhutan in the early months of 1989 were high-profile south Bhutanese men from both the most ritually pure and ritually polluting castes. This group of five took refuge in Nepal, hoping to garner international support for their cause, only to be extradited back to Bhutan in 1989 by the king of Nepal. By 1990, Nepal was swept up in a popular uprising that demanded the autocratic rule of the Royal Panchayat regime end and be replaced by a multiparty, democratic system (Pfaff-Czarnecka 1997, 443). In February 1991, the newly appointed Nepali prime minister officially allowed the Bhutanese refugees into the country, asserting that he hoped Bhutan would consider democratic reforms. This thinly veiled critique may have been a foray into realpolitik. Hoping to recover from the recently ended blockade from India, Nepal was struggling economically. A coup d'état suggested economic instability, potentially curtailing much needed international aid and investment. Highlighting the democratic product of the overthrow may have helped appease cautious international donors and trade partners. Economically, Bhutan and Nepal traded very little. Critiquing Bhutan to emphasize its own politically modern approach to governance functioned to cast the Nepali government in a favourable economic light.

Despite offering sanctuary, Nepal could not absorb, provide legal protection, or offer humanitarian assistance to what quickly escalated into a massive influx of refugees. Nepal lacked a legal framework for refugees to situate themselves in. Nepal's Citizenship Act of 1964 regulated refugees' legal status. This act permitted citizens of specific foreign countries who had relinquished their citizenship to gain Nepali citizenship. Bhutan was not one of these foreign countries. Thus, the Bhutanese refugees were considered aliens. Nepal was not a signatory to the Refugee Convention or the 1967 Protocol Relating to the Status of Refugees. As such, refugees could not access political rights and were not allowed to engage in economic activities or own certain kinds of property. Of course, being a signatory to the Refugee Convention and its protocol is not necessarily a guarantee of better protection, nor does not being a signatory automatically mean refugees will not be afforded basic human rights (Janmyr 2019). Nah (2016, 229) observed that states in South Asia "have preferred to negotiate over the protection of specific groups of refugees rather than grant asylum to all refugees in a systematic and impartial manner." This was evident in relation to

Nepal. Earlier flows of Tibetan refugees into Nepal were given protection, access to economic activities, and freedom of movement, despite the country not being a party to the international framework (Banki 2008a). By the early 1990s, both the regional dynamic and the political salutation in Nepal had shifted (Banki 2008a). As the number of refugees climbed and conditions deteriorated, the government of Nepal requested UNHCR assistance in September 1991 (Hutt 2003). Agreeing to host these refugees promised a positive means of differentiation from the earlier administration. Similar to other countries that host refugees, agreeing to allow the refugees to stay provided a pathway "to use their involvement in humanitarian responses…in attempts to improve their global standing" (Cevik and Sevin 2017, 399). Regardless of these potentially positive outcomes, the agreement to host the refugees had caveats. The group needed to be contained in camps, and repatriation to Bhutan was of the utmost importance. In this way, the refugees served an additional purpose. Nepal was reasserting its status as a nation-state with the ability to regulate its borders after an unsuccessful power play with India. Reflecting this, repatriating refugees to Bhutan became the singular solution sought by the Nepali government. Salter (2008, 370), building on Agamben (2005), articulates this stance:

the sovereign ability to define and limit the population is a longstanding institution of the state, intimately tied into the notion of sovereign territoriality and the imaginary of borders implied in this conception of bounded space.

Permitting refugees to stay in the far eastern region of Nepal reflects geographic, political, and economic concerns. The refugees were permitted to settle along a river the local community was not actively using. This location also reflects political decisions on the part of the Nepalese government: to keep the problem peripheral to the highly centralized political sphere of Kathmandu.

The first UNHCR-administered camp was established in April 1992, and six more were built over the next year. The flow of refugees steadily increased until an approximate peak of one thousand refugees per day was registered in the middle months of 1992 (Reilly 1994). As 1992 concluded, fifty-six thousand were under the care of the UNHCR.

Nepal Administrative Map, April 27, 2015

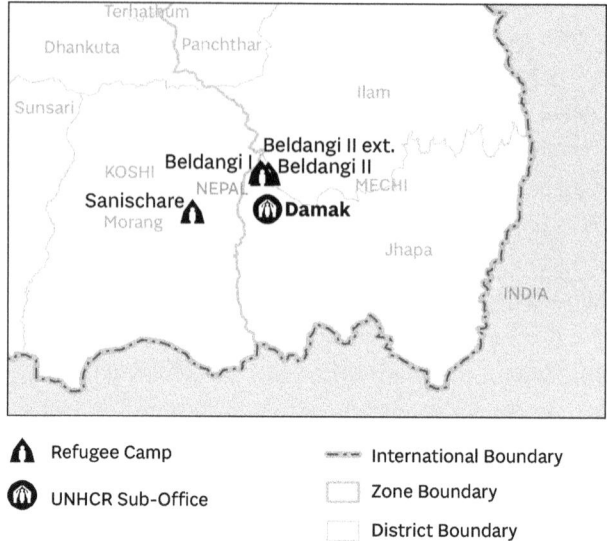

🛆 Refugee Camp

🏛 UNHCR Sub-Office

--- International Boundary

▭ Zone Boundary

▭ District Boundary

Nepal Administrative Map. (UNHCR, United Nations Office for the Coordination of Humanitarian Action, Nepal Survey Department 2015. Used with permission.)

This number increased to 75,400 in 1993, before the trickle of refugees slowed in 1996 and the camp population settled at eighty-eight thousand. By 2006, the year preceding resettlement, the total population was approximately 107,000 due largely to procreation (World Food Programme 2006). In addition to the camp dwellers, there were estimated to be tens of thousands of Bhutanese living in camp-like situations in India (US Committee for Refugees and Immigrants 2002).

The Beldangi camps were the largest and housed close to fifty thousand people by 1993. Beldangi consisted of three satellite camps—Beldangi I, II, and III (or Extension)—with a population greater than most cities in Nepal and roughly double the size of Thimphu, the capital of Bhutan (Reilly 1994). They were located approximately fifteen minutes north of the town of Damak (population 41,421 in 1991) (Dhakal and Strawn 1994; Brown 2001). These camps were created directly by the UNHCR, and their design reflected a need to house a high number of people while effectively delivering basic services. Thus, they were crowded. The camps sat on a floodplain that has been partially fortified by UNHCR-supplied sandbags. Fingers of scraggly jungle divided the camps, attracting the occasional elephant. Despite being established two decades previously, in 2012 and 2013 the camps still had a temporary feel.

The other four camps include Sanischare, Khudunabari, Timai, and Goldhap in the Jhapa and Morang districts (Banki 2008a). In 2012, due to the high rates of resettlement, all had been consolidated and closed except the Sanischare camp. Sanischare is markedly different from the Beldangi camps. Its construction, according to participants (I was not able to verify this claim), was supported directly by the European Union rather than the UNHCR. While Beldangi was remarkable for its congestion, Sanischare had a decidedly village feel. Despite the temporary nature of the bamboo housing, the space between the huts was greater. It was set on a hill, rather than a flood-prone plain, and this lent it an air of permanence. A small library and a few brick buildings accentuated these impressions. Rather than being some distance from town and buffered by jungle, the camp was adjacent to the bustling, small town of Patri. As the following chapters illustrate, these spatial configurations impacted on the experience of living within the camps.

Though the refugees were not governed by the Nepali state, the laws that functioned within the camps were developed in consultation between the government and the UNHCR. These laws reflected Nepali norms and values. Thus, an act such as playing cards was illegal both within and outside the camps. Though some laws created the impression of a shared system of governance, most laws reinforced the differences between refugees and citizens. The "camp policies and directives from the Government of Nepal restrict freedom of movement of refugees and prohibit their engagement in gainful activities, such as agriculture, trading, and other business" (World Food Programme 2006, 8). Technically, refugees were not able to work outside the camps, and the movement of refugees was theoretically highly regulated. Local integration was not a legal option for the Bhutanese refugees. Government decisions suggested a desire to differentiate between the two groups and keep Bhutanese refugees out of the local population: the message was that a Bhutanese refugee was different from a Nepalese citizen and needed to be contained. This was an exercise of power, an enforced segregation regarding who was included in the nation-state and who was excluded. The more ambivalent reality related to the peripheral location of the camps. Relegating camps to the fringe of the country served to buffer the highly centralized government from refugees' potential political demands. It also made their oversight difficult—if not impossible. On the edge of the nation of Nepal, the borders between camps and villages, Nepal and India, refugee and citizen were blurred. Regardless, the expectation of difference was persistent and became reinforced by the organizations working within the camps and refugee leaders. Further, refugee leaders had a stake in the establishment of camps maintained by the international community—international involvement was perhaps the only way their experiences and causes could gain credibility.

Conclusion

While the processes leading to the exile of close to one hundred thousand Bhutanese occurred in a seemingly remote part of the world, they are linked to broader regional and global contexts. Disputes over territory, migration, how to assimilate minority groups, and demographic anxieties precipitate exile in many contexts (Van Driem 1994). The events that occurred in Bhutan during the late 1980s to the mid-1990s are described as "ethnic tensions" (UNHCR 1992; Amnesty International 1992a; Human Rights Watch 2003). This chapter illustrates that this understanding of the refugee-generating process requires interrogation. Ethnic differences exist but became amplified by colonial discourses and local power struggles. The inequality that developed and festered in Bhutan was the product of competition over resources and, perhaps of equal significance, political power. Stereotypes buttress understandings of who claims political space in Bhutan. In Bhutan, radical ethnic stereotyping resulted in certain groups being pushed further to the periphery of state power. A very specific, local environment (interacting with a globally linked history) constructed them as marginalized, peripheral, and immoral: people on the fringe. Many of these understandings coloured global responses, positioning them as deserving subjects of humanitarianism.

The relationship between citizens and states is political: the state determines whom it includes and whom it excludes. If those excluded from states become a pressing, urgent humanitarian emergency, they may be evaluated as refugees. Once labelled, they enter a system of international governance headed by the UNHCR. As the UNHCR manages these populations, refugee camps become sites designed to address the causes of exile through social and ideological development. By normalizing this distinctly humanitarian development of refugees, complex conversations regarding global inequity are avoided. The refugees themselves, rather than the states that generate or perpetuate their exclusion, become legitimate populations for radical social intervention. Yet the Bhutanese are not completely passive in these situations: displacement can "be an opportunity to assert claims to political recognition" (Gatrell 2013, 283). However, contemporary approaches to refugees tend to obscure this political dimension, leading to a sanitized, apolitical community. The following chapters analyze the Bhutanese refugees' attempts to balance their unique value system with the norms and values of managing institutions.

3
Learning to Be Humanitarian Subjects

ARRIVING IN THE MAIN BELDANGI CAMPS, I was immediately struck by the congestion. Bamboo huts appeared to sprawl with mathematical precision in every direction. While striking, this was not surprising: images of crowded refugee camps have been transmitted around the globe for decades. The second impression was unexpected. Unlike the Syrian camps in Turkey, there were no high gates topped with barbed wire encircling the space. Occupants did not have to undergo retinal screening as in Malawi camps (Ghelli 2014), or a fingerprint check to enter or leave as in the Tanzanian camps (Ismail 2006). Identity cards did not appear to be checked with any kind of regularity, in contrast to camps such as Dadaab in Kenya (Horst 2006). Passes were not necessary to leave the camps, as in the Mishamo camps in Tanzania (Malkki 1995a).

These technologies all attempt to regulate the movement of refugees, effectively managing populations. In Nepal, refugees are managed quite differently. The physical division—walls, fences, regulated gates—between locals and refugees is less rigid than in other camp situations. A skeleton crew of local police was tasked with enforcement. Though there is a rhetoric of containment, the movement of people only had a semblance of moderation through the official or main entrance. It was unencumbered elsewhere. While the Beldangi camps were partially surrounded by barbed wire, the fencing seemed to start and stop without logic and had been co-opted for drying clothes, air-curing meat, or tethering animals. The UNHCR presence, aside from a logo on the occasional blanket, is minimal: UNHCR staff is based in nearby Damak and visit the camps infrequently. In place of guards or fences, refugees are inculcated with humanitarian values that position them as a deserving population. Thus, rather than physical containment and onsite surveillance, the emphasis in Nepal is governance from a distance, achieved through radical moral reconfiguration. Transforming the refugees into righteous, humanitarian subjects promises those excluded from the nation an opportunity to become part of a virtuous, international community (Rose 2000).

The refugee camps in Nepal are illustrative of the space between a humanitarian emergency and "the ordinary, longer-term problems that surround them" (Redfield 2013, 191). For the Bhutanese camps in Nepal, the problems that surround them include not only a global system that appears to perpetuate containment but also generates domestic power struggles. These domestic struggles occur between locals and refugees, particularly when local staff is employed by international organizations. Further, they occur within the group as multiple value systems intersect and compete for social authority. In other words, camps are far from passive relief zones. Rather, they are dynamic sites of social transformation. Malkki (1995a) observed this in Tanzania, arguing that camps foster essentialized social categories. Building on the premise that social categories are both made and unmade in the camp settings, this chapter examines the role of humanitarian governance as it attempts to mould refugees into particularly deserving subjects. As Ong (2003, 53) argues regarding Cambodian refugee camps in Thailand, after the emergency phase passes, "relief organisations quickly begin the task of physically and socially converting the refugees into citizens of the global world." These humanitarian subjects are tailored to reflect the legitimacy of the nation-state and the values sanctioned by the international organizations working within the camps. Camps become exceptional spaces, excluded from the nation but included in a supra-national network underpinned by dynamic, and at times fickle, humanitarian values. Internationally supported camps become sites of morally just intervention, aspiring to remake the world by designing righteous humanitarian subjects, but people within the camps continuously negotiate what "righteous" means to them.

Becoming Bhutanese

During the first two weeks of December 2012, there was one event on the minds of the refugees in the Beldangi camps: Bhutan's National Day. The National Day falls on December 17 and commemorates the coronation of the first king of Bhutan, Ugyen Jigme Wangchuck. Bhutan always is a popular topic of conversation in the camps, but the impending date made it the focus of even more conversations. Speculation was rife: Would the king pardon prisoners? What would the king discuss in his speech? Would the refugees get a mention? The appropriate way to celebrate what I was repeatedly told was a very important occasion was discussed at length. Deciding the exact activities seemed to be the only problematic aspect. There were assertions that there should be dart throwing or a soccer game. Other people surmised that there would be dancing and music—requisites of any celebration. Everyone assured me many festivities would be taking place. Those who owned a *gho* or *kira*—Bhutan's national dress—would don them, and perhaps the children would put on a special demonstration of Dzongkha, Bhutan's national language. The day, like other festival days, would start early and stretch throughout the evening. It was not to be missed.

The lead-up to Bhutan's National Day reflected the first lesson the refugees learned in the camps: the importance of being Bhutanese. In essence, the refugees were only entitled to support if they were outside their nation-state, and, crucially, they must desire to return home. Such values were taught via an "education for repatriation," both formally in the camps' schools and informally through institutional programs in the camps (Evans 2010b). These projects were undertaken with the hope of being allowed to return to Bhutan—a reflection of the goals of some refugees, the host government, and the UNHCR. Educational endeavours included promoting the acquisition of Bhutan's national language and history. Additionally, the few south Bhutanese who had lived in the north of Bhutan taught Drukpa etiquette to the broader group, which included how to greet a superior, the appropriate way to eat out of a bowl, and how to cross one's legs when sitting. In short, the refugees needed to embody the physical demeanour of a particular kind of Bhutanese.

While living in the camps, several of the refugees purchased Bhutan's national dress, while others hung photos of the king in their huts. While Bhutan stipulated that speaking Dzongkha and being able to exhibit the correct social etiquette were key factors in being a Bhutanese citizen, Bhutan gave no indication that the camp-based efforts would impact on the chance of repatriation. The government of Bhutan maintains an unwavering stance that it holds *no* responsibility towards the refugees. Most participants, rather than question the push for repatriation or protest the imposition of a largely alien culture, described this process as a time of learning. They learned from the international organizations working in the camps that these transformations were legitimate requests by nation-states. Nevertheless, a few participants mentioned how disconcerting it was to promote the customs of a country they had been forced to flee. While the managing institutions and the host government promoted the idea that the refugees naturally belonged in Bhutan, this understanding overlooked the myriad complicating factors hindering possible return. Further, for refugees who experienced imprisonment, torture, or assault sanctioned by the government, repatriation was not an ideal solution. These efforts, undertaken with the intention of helping refugees, attempted to transform the refugees into well-behaved citizens. In doing so, they illustrated that "the function of modern international organisations is to manage refugee populations in a manner that does not radically undermine the framework on which the nation-state rests" (Owens 2009, 571).

The dominant message to emerge from these educational endeavours was the significance of national affiliations. Focusing on the national aspect of the group, rather than the numerous ethnic and linguistic divisions, allowed the managing institutions to approach the group as a nation in miniature. No longer a stream of refugees, the group became a containable entity. The UNHCR does not explicitly define *community* in its numerous handbooks, but in Nepal the organization implicitly conflated community with place. This generalization very effectively created a Bhutanese community, bound in a similar fashion as nation-states. Quite quickly, the group was approached bureaucratically as a

Bhutanese refugee community. As a community, its governance could be expedited. Activities were delegated to community leaders, who were tasked with the moral self-regulation of the group.

The dual statuses of Bhutanese and refugee became intrinsically linked. In turn, these associations legitimized the international support they received in the camps. Their exile entitled them to particular amenities to which the host population was theoretically not entitled. The refugees were acutely aware that it was only through their exclusion by the nation (not only Bhutan but also Nepal) that they could access any of the security that the international community provided in the camps. This inclusion was conditional on the refugees' willingness to maintain the largely artificially defined boundary of "Bhutanese refugee community" created in the camps, a label that reflected a broader humanitarian ethic. The refugees necessarily emphasized the line between local and refugee to me. During fieldwork, resettlement was well underway, and many refugees were involved in processes that evaluated if they were eligible for resettlement. Though I was forthcoming that my presence had no bearing on their resettlement outcome, there may have been a sense they needed to present the "right" image of a Bhutanese refugee to a relative outsider with familiar and citizenship ties to the United States and Australia. The power dynamic between the participants and myself certainly coloured their response to my presence in the camps. However, because of my interest in their presentation of self in relation to local and global forms of governance, this was not always detrimental to my research.

Bhutanese Are Righteous

Resources available to the refugees were, in many respects, rudimentary. Electricity was nonexistent, and fires from oil lamps or wood cooking frequently broke out, spreading with frightening speed through the cramped quarters. The bamboo huts were bitterly cold in winter. However, the belief persisted that camp residents were receiving better treatment than the local population in terms of food allocation, access to medical care, and educational opportunities—a perception validated by the significantly lower rates of malnutrition, under-five mortality, and incidents of infectious disease in relation to Nepal's statistics (Brown 2001). The refugee camps (and in particular the large Beldangi camps) strained the local environment in terms of land use and made access to forest resources more competitive. This contributed to growing tensions between the refugee group and the local population (Birendra and Nagata 2006; Banki 2008b). Though the refugees were technically not able to work outside the camps, and the movement of refugees was in theory highly regulated, thousands of refugees moved between the camps and the town of Damak each day. Though some members of the local community benefitted from an invisible workforce, the refugees were also viewed as unfair competition for local jobs.

It is not surprising that the relationship between locals and refugees is dynamic. Host communities may at first feel sympathy for refugees, but this can change when a timeline for return is not obvious (Ghosn et al. 2019). Further, when refugees have access to resources like education, health care, and food stability, while the local community does not, this can lead to tense relationships (UNHCR 2019, 24). In response to these tensions, the UNHCR implemented projects to benefit the host population, including improving local infrastructure, training of local medical care providers, and allowing local children into the refugee schools (UNHCR 2009). Roads leading to the camps were paved and buildings constructed that would be turned over to the community after resettlement was completed. The UNHCR strove to hire local Nepalese as a means of investing in the host community and did not hire Bhutanese refugees. Local Nepalese were the dominant face of the UNHCR in the camps. The refugee population was thus both a burden and a benefit for the local population. The balance between these two perceptions was dynamic and impacted group interactions. One participant articulated,

> Locally, internally they do not looking good upon us. Externally, they are looking good on us because we give them business. The good terms are on the outside but the inside is bad. If they want something from us, then they are good to us. We are not on good terms with our family in Nepal because they were very selfish and would not help us when we came from Bhutan. Now they say, "when your process for resettlement is started they will send us some money." Their view is that this will help them; we'll send them money when we resettle (Chhetri male, age 33, interview, Beldangi, 2012).

Another man explained,

The local population demand many things from the UNHCR; *"the road is dusty, sprinkle it with water. The refugees are deforesting, reimburse us. We want, we want." This is obviously ridiculous, they have big houses and rice paddies. They use these words talking to us, "refee," "bhutangi" and Baganda (unknown caste male, unknown age, interview, Beldangi, 2012).*

Participants consistently articulated that, when the refugees are viewed as bringing beneficial projects to the region or money through remittances, they are treated well. When they are viewed as competition in the labour market, degrading the local forest (Birendra and Nagata 2006), or receiving more than the local population in terms of valuable resources (Brown 2001), treatment deteriorates. This varied though, with some refugees and locals enjoying an amicable relationship.

The closest hill tribe villages northeast of the Beldangi camps were poor relative to those living closer to Damak yet had good relations with the refugees. Because animals were not allowed in the camp, many refugees paid the villagers to allow them to keep the animals on their land. They often paid the villagers in rice (rations). The villagers were happy with this arrangement and would often sell firewood to the refugees, as this was illegal for the refugees to gather. When I visited these villages with refugee participants, we were greeted warmly. A participant explained that this group is better to the refugees because "they were originally very poor and have established a good relationship with the refugees. The people on the other side of the camps were very rich to begin with and that is why they are often jealous of things the refugees receive" (Brahmin male, age mid-20s, interview, Beldangi, 2013). These shifting and variable interactions illustrate the complexity of these relationships. The image of a Bhutanese refugee is linked to the manner in which the local group views them; however, this process of creating boundaries around identities is far from static (Barth 1969).

Attempts to create, solidify, and maintain divisions between the refugee and local population reinforced nation-based social constructs. Participants recognized this was difficult to maintain in the camps due to the shared lingua franca of Nepali, common ethnic backgrounds, and religions. The two groups spoke Nepali with very little, if any, difference in accents. Locals described the Bhutanese as "speaking a little old fashion" (local Nepali female, age late 30s, interview, Damak, October 2012). However, when I travelled outside the camps with the Bhutanese, locals did not assume they were refugees based on their accent. When I asked them how they knew the person I was with was a refugee, they asserted that I was spending time with them, and international people in Damak are there for the refugees. Refugee participants told me that in the past refugees were not able to walk with their heads up (with pride) in town. This has changed, and there did not appear to be a difference in the posture of locals and refugees. The ethnic groups found in the camps are found in greater Nepal. During religious rituals marking birth, marriage, and death, differences were virtually indistinguishable. The festivals that dotted the Hindu religious calendar were celebrated both by locals and refugees in nearly identical fashions.

Beyond that, in the mundane every day, the groups were remarkably similar. Refugees and locals washed clothes in a similar manner, hanging all items sopping wet on a clothesline so they dried without wrinkles. Newspapers were recycled to line tables or countertops. Locals and refugees, when eating dry foods such as puffed rice, tossed the food in such a manner that their hands did not make contact with their mouths. The kitchen gardens were similarly cultivated, punctuated with marigolds, enormous daikon radishes, and scraggly okra. The bounty of green, chicory-like *saag* was preserved by refugees and locals in identical fashion through a process of repeatedly dousing it with hot water and pounding it with a heavy, pestle-like stick before being left to ferment. Lower castes added dried fish, while the higher castes did not—irrespective of national affiliation. In terms of everyday life, the much-discussed line between Bhutanese and Nepalese appeared vague. Yet this line was essential to maintain, for it was on this line that political claims rested, camp livelihood depended, and access to resettlement depended. Thus, the maintenance of this seemingly irrevocable difference was significant.

The division between the groups became a matter of ideology and morals—without it, the assistance refugees received was called into question. The Bhutanese were educated; the Nepalese were backwards. The Bhutanese were righteous; the Nepalese were corrupt. Several participants used derogatory terms to describe the locals: "they are no good," "they are rotten," or "they don't have pure hearts." These comments came from all ages, both genders, and virtually every caste—particularly when I first began work in the camps. They were especially critical when describing locals who worked for international organizations. In contrast, expatriate staff were described as "so good" and "having pure hearts." The refugees were quick to associate themselves with morally righteous international actors who were only occasional visitors in the camps. They were acutely aware that, though largely unseen, the international community (rather than their local Nepali representatives they saw on a daily basis) controlled the camps.

Most participants described the moral difference between locals and refugees as stemming from the degree of education the refugees attained. In Nepal, education took place in the camp classrooms. The schools in the camps were drastically different from those in Bhutan. Caritas–Nepal is one of the UNHCR's implementing partners in the camps. It is a Catholic charity that strives to empower vulnerable populations through a variety of programs, including access to education (Caritas–Nepal 2021). The Caritas-run education system in the camps provided free, English-based classes, and child attendance was mandatory up to year ten for both genders. With textbooks and supplies provided to the refugees, the completion rate was close to 100 per cent at the primary and secondary level. In broader Nepal, primary schools also had high rates of enrolment, but supplies were difficult to procure. In terms of overall literacy, 75 per cent of the Bhutanese in the camps were literate compared to 57 per cent of the population of Nepal (Muggah 2005, Table 1). The camp-based schools explicitly promoted gender equality, the rights of children, and international human rights. In this school setting, the caste system was described as a human rights violation and was systematically undermined: higher-caste and low-caste children were encouraged to sit and learn together as equals.

For a brief period, schools provided a cooked meal for the students. According to one participant, in the Beldangi camps higher-caste students were obliged to eat food served by lower-caste students or teachers. He recalled trying to negotiate his family's expectations with the experiences in school:

> used to cook big pot of food and class captain would distribute the food but if it came from a lower caste, we wouldn't eat. We asked to change the captain, we want to eat but he's distributing. If I eat my mother and father won't allow inside home. Sometimes I was punished for asking this: I got the stick twice for this. I complained and the teacher from a lower caste so he punished [me] (Brahmin male, age 24, interview, Beldangi, 2013).

A refusal to eat this food, considered ritually polluting, could lead to corporal punishment, taunts by fellow students, and ostracism, while eating the food led to tensions in the family home. Attendance for children was mandatory, and participants told me that parents could be punished with cuts to rations if their children did not attend. Though there was a brief period of food distribution at schools, I was not able to verify the policy of withholding rations through the UNHCR or Caritas documents. It seems unlikely that the forced consumption of food was a top-down policy decision. I was also not able to verify that this occurred at the Sanischare camp, which suggests that, when this did occur, it was over a fairly brief period and isolated to the Beldangi camps. Yet I had five separate participants relay these experiences, suggesting at least one teacher may have imposed corporal punishment for refusing to eat across caste lines. The impression that rations could be withheld for not behaving did seem to be linked to the school experience and was, at the very least, an influential rumour. This rumour—where high castes could be punished and have rations withheld for not modifying their value system to reflect the governing organizations' ideals—pushed one powerful moral hierarchy (the caste system) against another (humanitarian: all lives are morally equivalent). In this context, only particularly "good" subjects that reflect the right values became legitimate recipients of humanitarian assistance. Refugees became different from locals not just because of the number of years of schooling completed but also as the result of a radically different value system that was reinforced in the camps.

The refugees prided themselves on being educated, and this education was based on specific social values. The experience of exile and coming under the care of international organizations led several to claim that they now understood the world—and their associated place—in a profoundly different manner. As one male refugee (unknown age, interview, Beldangi, 2013) explained, "From our camp, we have done Class 12. Before, I heard about refugees in other countries and didn't know who they were. Now I am one and I understand." Similarly, a female refugee in her early thirties stated, "Education began occurring at a higher level (leading to) a social awareness. It did a great role in equalizing, the equality of the people" (interview, Beldangi, 2012). This statement accentuated a key aspect of the camps: the need to reflect the values that were promoted in the camps by international organizations.

When I first began talking to people about their educational experiences in Nepal, they articulated that the Nepali-speaking Bhutanese have always been highly educated: more educated than the Bhutanese elites that expelled them. Several speculated that the reason the government expelled them was because they were highly educated: "We were skilled, brilliant people and they became jealous. That is why we're refugees" (Brahmin male, unknown age, interview, Beldangi, 2012). Another also articulated this understanding: "Drukpa saw that we were so skilled and educated, they worried the kingdom would be harmed, this is the reason we became refugees" (Brahmin male, age 40s, interview, Sanischare, 2012). Many of those that pursued education beyond year ten during the late 1980s did so in India, and a few refugees were able to obtain an advanced degree outside the camps in Indian schools (Brahmin males, ages 20s–late 40s, interviews, Beldangi, 2013). I cannot recall a single participant that did not hold a high degree of education in esteem, yet while high-caste men were eager to discuss their experiences studying in Bhutan and India, these stories did not emerge when talking to other people in the camps. I began to suspect that education in Bhutan was an exclusive activity.

I quickly realized, as I spoke to women and members of the lower castes, that the community leaders' (Brahmins, and Chhetris to a lesser degree) experience of education was not representative of the broader community. When I arrived in Nepal, an introduction via my contacts in Australia resulted in a detailed list of people whom I was "supposed" to speak with while in the camps. This was incredibly helpful, but I quickly realized that the list was problematic. The selection of a research population by "community leaders" has been critiqued for failing to recognize intergroup power struggles (Mackenzie et al. 2007). Participants may have "their own biased understanding of community belonging...it is vital to acknowledge how implicit imbalances of power" can impact on research (de Smet et al. 2021, 10). The list I was provided was far from representative. Very few women were on the list and, though perhaps a few low-caste men, no low-caste women.

To ensure data was relatively representative of the broader group, in addition to seeking participants independently of my research assistant, I developed a survey. Heeding the advice of Bryman (1988), who argues surveys should be developed in consultation, this survey emerged after five months of fieldwork in consultation with my research assistant. My research assistant and I spent approximately two weeks conducting the initial twenty-five surveys, asking for feedback from participants and honing the final questions. We elected to survey 1 per cent of the Beldangi II camp population of fifteen thousand and sought to have those 150 surveys be demographically representative of the camp population. The sample group was based on age, gender, and caste or ethnicity. For example, there were approximately 1,600 Gurung[1] in the Beldangi camps, thus we would survey sixteen people—eight males and eight females. This would be further broken down by age. Of course, there were limits to surveying only 1 per cent of the population, and a larger sample size would have been ideal.

In many regards, the survey's most important function was as means of introduction to people deemed by those trying to direct my research as inappropriate. Thus, I used the camp-based survey primarily as a means rather than an end. Yet the surveys ultimately provided a robust body of demographic statistics demarked by age, gender, ethnicity, and caste. In particular, the data was useful in exploring historic educational trends in relation to caste and gender. Based on my surveys, of women who grew up entirely in Bhutan, 7.5 per cent attended school

---

1. Caste and ethnicity are often directly related in the group. Participants who have converted either to Christianity, Buddhism, or an animist religion are still well aware of their caste. It is important to note that, in the eyes of the high-caste, Nepali-speaking Bhutanese, the caste that a person is born into can never change. Regardless of educational attainment or enjoying high economic status, there are not mechanisms in place to "move up" the social ladder. The terms "caste" and "surname" are often used interchangeably by participants.

in Bhutan. Those who attended school stopped their education once menstruation began—generally before year seven. The average grade completed by women in Bhutan was year four. In conversation, several women recalled that, in Bhutan, "women were not allowed to join school for this was the community belief" (females, ages 30-70, interviews, Beldangi, Sanischare). Brahmin women articulated that when they were children and young women in Bhutan, an educated woman was considered "a curse." For men raised in Bhutan, approximately 25 per cent attended school. The median grade completed for this group was year five. Brahmin males achieved the highest level of schooling, studying up to year ten in Sanskrit schools. They also were disproportionately better educated, with 60 per cent completing some level of schooling in Bhutan. Thus, while the theme of the refugees being highly educated while living in Bhutan was incredibly widespread, surveys revealed that virtually no castes aside from the Brahmin males attended any kind of schooling in Bhutan. This finding began to illustrate how a particular story or experience had come to dominate within the larger group. While my focus was not to discern the facts behind the group's myths, I was interested in exploring why they developed and how they were regulated. The survey data further illustrated the centrality of Brahmin males in public representations of the "Bhutanese refugee community."

Transforming Behaviours and Internalizing Values
The Bhutanese actively promoted themselves as socially aware and embodying the value of equality. This understanding of equality was set in contrast to the local population, particularly when discussing the caste system. The caste system in the camps was explained to outsiders as an unfortunate consequence of living in close proximity to the Nepalese. The system allegedly existed in the camps due to the corrupting influence of the locals, but it was fading due to, as a female in her thirties suggested, "the higher education and this means greater equality, also there are many programs running in the camps" (interview, Beldangi, 2013). While there were several types of programs running in the camps, they could broadly be divided into two categories: those that attempted to transform behaviour, and those that illustrated the transformations that had occurred. The former were largely dictated to the refugees by international humanitarian organizations in order to teach

morally righteous values. An example of this was the way camp cleaning duties were allocated. Billboards punctuated the camps with images of men washing dishes or fetching water, an attempt by the international organizations to change gender roles. There was also a clear expectation that everyone would clean the camps. This was also a radical break from the historic system that demanded only particular low castes were responsible for cleaning outside the home. Camp cleaning rosters were posted and strictly adhered to. These rosters included all ages, male and female, and every caste; everyone was expected to keep a clean camp for the sake of their community. Unprompted, a participant brought me to the roster, explaining the changes that occurred in the community. He explained that though there was still some resistance to these changes, the refugees themselves could address them and that they had developed organizations within the camp to do so: they no longer needed outside interventions via posters or training programs. They wanted the opportunity for refugees to teach each other these values. By demonstrating that they had internalized these values, the refugees created an image of being domesticated (controlled) subjects eligible for ongoing care and support. The ability often hinged on the provision of financial support but also reflected a deep desire for autonomy. This echoed the

> *new emphasis on the personal responsibilities of individuals, their families and their communities for their own future well-being and upon their own obligation to take active steps to secure this (Rose 1996, 327–328).*

A second group of projects was particularly illuminating regarding the ways refugees were transformed into humanitarian subjects. In Bhutanese culture, fertility is celebrated: a woman's red sari is a bold reminder of her reproductive capacities. Children are highly valued for social, economic, and religious reasons (Stone 1978). Historically in Bhutan, children participated in rice harvests thereby contributing to the economic viability of the family. Families were large: older women claimed upwards of seven children and a few boasted fourteen. Even in the camps, having only one child, even if he was a male, was still considered inauspicious. As several women informed me, "One is not good, what if something happened? Who would take care of you? One boy is not enough, it is better to have a girl as well—they can help you around the house" (females, ages 20–70, interviews, Beldangi and Sanischare).

The refugees I worked with loved their children, and it was evident that children brought great joy not only to families but also the broader community. In addition to the joy that children can provide, children represent a social security net for women and for many were the key path to social status. Nevertheless, when I was in the camps in 2012, there was a campaign to introduce wider use of contraceptives. A smaller family size was promoted as being more cost-effective, aligning with the shift within UNHCR from an emphasis on relief to one on development. The people to whom I spoke agreed with the premise of these campaigns. Men and women both explained that, with such uncertainty about the future, maybe a smaller family would be better. Others questioned me directly about what the acceptable family size was in either the United States or Australia. Some worried that a large family would become penalized. Not a single participant directly stated they would not participate in the contraception education programs running in the camps. Rather, they saw their participation as a crucial means of showing their support for the UNHCR policy and broader ideals of family composition.

Most efforts were largely externally designed, developed, and dictated, with the goal of promoting equality in the camps. However, there was also a strong push for the camps to be self-sufficient and self-reliant. The hope was that refugees would undertake these types of efforts of their own accord. The goal was to create a community that could self-regulate. "Ethical values, rather than direct surveillance, became a means to maintain social order" (Neikirk 2018, 71). This aspiration was illustrated by a series of workshops run by an international NGO working in the camps with the permission of the UNHCR. These workshops occurred the month before I began working in the camps but were frequently discussed by participants. A twenty-seven-year-old refugee provided a succinct summary of one of the workshops:

He had us all sit and imagine we were in a room. The room did not have doors or windows. Now we sat in this room but we were unhappy and we wanted out. We looked and looked for a door but there wasn't one. It was frustrating. He told us we could be stuck like this forever: looking and looking for a door or a window. Then, he said we had to imagine we were looking up. Instead of a ceiling it was just open. We could leave the whole time...What do you think about that? We are not too sure (Brahmin male, interview, Nepal, 2013).

The exercise is aimed at "creating active individuals who will take responsibilities for their own fates" (Rose 2000, 337). Rose (1999, 2000) is referring to governmentality in advanced liberal democracies: techniques of social control that shift responsibility away from states while promoting the idea of personal empowerment: we choose our fate. In some instances, like the training activity discussed above, one can almost imagine this activity being run at an employee training session in the United States or Australia. My argument is that these norms, values, and ideals are being transplanted into refugee camps. Rather than questioning why they were trapped in a hut or strategizing how they could seek assistance, their problems became an inability to help themselves or develop innovative solutions. While promoting the goal of self-sufficiency, this lesson also functioned to regulate the group. This regulation was achieved by introducing a degree of conditionality: the refugees must conduct themselves appropriately. They must actively work towards improving their quality of life rather than, to use the analogy of the NGO, feel frustrated by an apparent lack of doors. By "looking up" to the humanitarian values espoused in the camps, their situation would be improved. As Ong (2003) similarly observed in the Thai camps that host Cambodian refugees, the emphasis was on creating self-reliant, disciplined, and controlled subjects. In both contexts, this lesson was highly individualistic while also reinforcing the ideals that individuals should undertake projects to better their community.

The multiple participants who discussed this workshop with me proceeded to ask for my additional perspective as a representative of a broader international network. The refugees were trying to determine my values as a member of their audience. This demonstrated that, "surrounded by a web of vocabularies, injunctions, promises, dire warnings and threats of interventions, organised increasingly around a proliferation of norms and normative," the humanitarian subject must constantly self-evaluate (Rose 2006, 150). Repeatedly, the refugees had to reinvent themselves and mirror continually morphing expectations of humanitarian agencies in the camps. The need to reflect ideals and expectations thus underpinned the drastic power imbalance between those who worked and those who lived within the camp. The questions participants asked regarding this exercise were attempts to stake a small claim over the various activities they were expected to accept or embrace in their entirety. These projects were rarely partnerships, even if undertaken with the best of intentions.

However, there were activities theoretically developed and controlled by refugees. These projects hinted at the possibility that if the appropriate values were internalized, then the group would be rewarded. In the camps there was a glut of projects undertaken by refugees all striving towards social transformation. One participant, raised in the camps, summarized the driving ideal: "development starts with us" (Rai female, age 20s, interview, Nepal, 2012). These projects were run, as multiple ambitious coordinators explained, "to support our community." Participants stressed that serving their community was a good or righteous activity. This mirrored the aspirations of the governing institutions. Reflecting these values translated into various privileges (Rose 2000). Though the refugees were expected to volunteer their time, in order for these projects to take place, they required funding by either the UNHCR or another NGO working in the camps. Funding generally went to constructing a hut to host activities, defraying printing costs, or procuring specific materials. The lack of funding for wages, far from being criticized, was often praised for giving the refugees the opportunity to volunteer. As a female refugee explained, "Why do I volunteer? It feels that work as a volunteer is a good job. It is done without profit so it is good" (Bhujel female, age 20s, interview, Nepal, 2013).

The UNHCR stressed that paying refugees a wage for performing community services would lead not only to corruption but social deterioration (UNHCR 2008). The Bhutanese, far from criticizing this decision, lauded it as a way of protecting them from the corrupting influence of paid community service. Yet it did mean that some had to work in more precarious situations if their family required an injection of cash. However, in the camps they presented themselves to representatives of international organizations, and to researchers such as myself, as proxies for the international organizations, tirelessly working to reach the prescribed level of moral values. This contrasted with Nepalese citizens who occupied paid positions in the UNHCR, IOM, and most other organizations in the camps, who wanted to give a local face to their activities. This division again reinforced the divide between local and refugee. The refugees were serving their community, while the locals were perceived as profiting from their exile. Maintaining that refugees help each other due to moral righteousness (versus economic gain) ensured ongoing access to resources not available to the host population. In turn, this fostered a sense of conditional inclusion in a broader, international framework.

Projects focused on helping vulnerable people were viewed as consistently receiving funding and support. "Vulnerable" is a term employed by the UNHCR to denote groups that require additional social and/or economic support. The UNHCR defines *vulnerable people* as women (particularly single), children, the elderly, mentally or physically disabled, ethnic minorities, survivors of violence, and those suffering from HIV/AIDS (UNHCR 2005). To borrow the language of an idealistic young refugee, "We need to help the poor people—you know, the victims" (Brahmin male, age late 20s, interview, Nepal, 2013). Helping the "poor people" was an act that had roots in Hindu cosmology. The giving of charitable alms to the poor is promoted in the Vedas (Bornstein 2012). However, most participants felt the charity work in the camp was different. Religiously, charitable giving could contribute to one's karmic merit—the rewards were not necessarily recognized in this life. But in the camps, participating in a charitable framework promised resources and recognition. Training programs yielded certificates that had potential value once resettled. Service to one's community enhanced the image of the refugees in the eyes of international actors and donors.

This is not to say there was not altruistic giving occurring in the camps; I observed women cleaning the temples without expectation of recognition, and goods or services provided to those that needed them. Rather, the structured programs in the camps began to mirror the governing institutions. The newly created Bhutanese refugee community became tasked with identifying those in need. Those who possessed the correct qualifications and values intervened with the support of international staff. They became the gatekeepers to the group and in doing so had considerable sway over how the community was represented. These tended to be people who could readily use the international discourse and nomenclature. This was strategic: "those who learn to translate their needs into the language of others may find valuable resources and support" (Eversole 2010, 37). Hence, "vulnerable," "community service," and "development" punctuated discussions regarding programs in the camps. All these buzzwords highlighted the necessity of intervention: a moral evaluation of the refugees had found them wanting.

Transformational Projects

It appeared that projects were funded on their potential to transform refugees. This underscored the understanding that the group required intervention and needed to change. The projects that developed in the camps to help vulnerable groups included children's centres, domestic violence support centres, designated classrooms for the disabled integrated with education strategies tailored to their unique needs, and literacy classes for the elderly. These interventions, though perhaps well meaning, were often based upon assumptions and understandings that did not necessarily reflect the experiences of refugees. The much-praised domestic violence program in the camps illustrated this quite clearly. Female refugees, trained by international organizations, staffed the domestic violence centres in the camps, called Gender Focal Point. Once trained, staff ran educational programs in the camps to promote gender equality, the dangers of alcohol, and the impact of domestic violence on children. They published booklets, held mediation sessions, and provided counselling for families. The central message focused on the equality between men and women as a basic human right. The coordinator, a twenty-six-year-old woman, was at pains to highlight the effectiveness of the centre's activities. There were less referrals and fewer reports of domestic violence between men and women. This outcome was described as a very good thing—"violence has been controlled through campaigning and programs, violence is decreasing" (Bhujel female, interview, Nepal, 2012, 2013). We spoke at length regarding her training, the violence in families, the support systems in place, and why she thought there was violence in the camp. She kept reiterating the successes of the program, reminding me again that the rates of violence between husband and wife had dropped over the past year.

In terms of effectiveness (a crucial factor in maintaining funding), the program seemed to be working. It also felt right. The centre was run by refugees and headed by women. In many ways, the centre seemed the perfect manifestation of camp governance: through guidance and management by international donors, the refugees internalized the necessary ideals and learned to be "good." I was feeling the interview might be wrapping up when she stated, "You know, it is not just husbands and wives, the worst kinds of violence are between women" (Bhujel female, interview, Nepal, 2012, 2013). Intra-female violence was a significant and continuing problem in the camps but did not receive the same attention as spousal violence. This limited understanding ultimately hindered attempts by victims of other forms of domestic violence to access resources and support in the camps. The Bhutanese are strongly patrilocal; a new wife lives with her husband's family and contributes considerably to the household. Younger women slowly gain social status as they bear sons and grow older. For women, their status reaches its pinnacle once the eldest son marries and brings his wife into the family home. This status can remain high even if the son's mother is widowed. These senior women, if living with their son (a standard arrangement in the camps), can wield considerable power.

The new wife enters this household power dynamic. Immediately, she is expected to contribute in several ways. Under the guidance of her mother-in-law, she performs the bulk of domestic duties. Before dawn, she fetches water to ensure it is warm or tea has been made when the family rises. The responsibility for cooking meals and snacks, processing milk into yoghurt and ghee, along with cleaning and washing, falls to her. The garden has to be tended, and every few months the hut's floor needs to be refinished with cow dung. For most women in the camps, the only reprieve from the workday comes in the late afternoon—a time for neighbourhood gossip, crocheting, and knitting. This arrangement is maintained until the bride becomes pregnant with her first child. Then the mother-in-law runs the household while the expectant mother first convalesces, then, after childbirth, tends to the child. This arrangement can last up to a year. After this period of "maternity leave," the process of accumulating domestic credit begins again.

This was the idealized situation, but it did not always work. There could be personality conflicts, different tastes in food, or distinctive standards of cleanliness, leading to tension in the household. There could also be conflicts regarding when maternal rest should begin or end, or the foods the infant's mother, dependent on her mother-in-law to prepare, wanted to eat. This tension could erupt violently, generally with senior women physically assaulting junior women.

Such a form of domestic violence did not fit into an imagining of domestic violence that the project funders transmitted to the refugee staff. From the project's standpoint, elderly females who had outlived their husbands were considered vulnerable. For a junior woman, who was married and theoretically not vulnerable, to experience physical abuse by a "vulnerable" elderly woman complicated access to support. The elderly woman (the mother-in-law) was entitled to additional resources from the UNHCR and its implementing partners. At a household level, there was reluctance to identify the mother-in-law as anything other than vulnerable for fear of losing additional resources. This situation illustrated the limits of a generalized or collectivized understanding of women refugees, both elderly and vulnerable. Such templates for approaching life in the camps began to strain under diverse social arrangements.

The coordinator did not collect data regarding the incidents of female–female violence but considered it to be an equal, if not greater, issue in the camps. Assaults between women sharing a home were not a focus of the international training sessions and did not become translated into the camps through campaigns or programs. The coordinator of the domestic violence centre felt compelled by international overseers to focus primarily on male–female attacks. Her successful mediation of these situations evidenced the group's ability to self-regulate. Focusing on this particular kind of domestic violence, which was decreasing considerably within the camps, promoted a positive image of the Bhutanese. These measurable successes then reinforced the framework the program operated within: the program worked because the refugees were effectively helping themselves.

The refugees were, in effect, created in a particular image requiring corresponding interventions. This project received funding because it ticked the right boxes: female refugees running a gender centre to stamp out male-perpetrated domestic violence. A template of who was vulnerable was projected onto the group, yet the reality was perhaps inevitably more complex. In turn, this placed the coordinator in a conundrum: to maintain funding and the centre, she needed to focus on the domestic violence that was targeted by the program. The difficulty in explaining to her funders that domestic violence was occurring between women was compounded because it could be constructed as yet another problem requiring external evaluation, intervention, and rectification. This illustrated that particular administrative imaginings of social arrangements must be mirrored to access resources and maintain a positive image. The requisite mirroring of particular norms and values to maintain social support reflects that refugees are well aware they cannot directly claim the rights of equality, self-reliance, and community support the managing institutions promote.

While the projects that received funding drove the values that were transmitted in the camps, those that were not funded were also significant. One male participant, after being unsuccessful at receiving funding for his fledging social service organization, commented to me, "They [international organizations] always are focusing on the negative, they also need to pay attention to what we do right. They say they want refugees to help-self but then do not give funding" (Magar male, age 40s, interview, Nepal, 2013). While this reflected personal frustration regarding the conditional status of funding in the camps, it was also indicative of broader camp processes. According to the participants, newspapers were not funded by the UNHCR. Refugees, with intermittent support from international organizations not working directly in the camps, largely funded the newspaper running in the camps between 2012 and 2013. Similarly, a refugee organization, the Delta group, which wanted to create an archive of the documents people carried with them from Bhutan, was not able to secure funding. Regardless of the reasons given for the lack of funding, the representatives of these groups took away the lesson that overly political actions, notwithstanding the UNHCR's emphasis on developing democratic management structures, were not rewarded. In turn, this underscored the broader, institutional shift away from the political aspects of exile. Rather, the focus in the camps became the creation of humanitarian subjects.

For the Bhutanese, subjective and evaluative project funding illustrated key expected refugee attributes, for example, the importance of being linked to Bhutan while looking towards the future. Rather than attempting to archive past grievances, they felt they should be bettering themselves in line with humanitarian values. It was through this process of becoming humanitarian subjects that "new spaces of visibility" appeared (Feldman 2008, 500). Gabiam (2012) similarly made this observation while working in Palestinian camps. For these Palestinians, abject suffering via living in slum-like conditions was equated with international support (Gabiam 2012). Development projects, in the Palestinian context, were viewed with suspicion: improvements to living conditions could undermine their cultivated political image. The striking difference between the Bhutanese and the Palestinian refugees was that the Palestinians saw the development of adequate housing as a direct threat to their status as deserving refugees. For Palestinians, the loss of "deserving" status was viewed as undermining efforts to return to Palestine. The Bhutanese, on the other hand, experienced a very different kind of development program. The possibility of improving the housing in the camps was not an option, thus the projects they participated in focused on improving their morality rather than their material situation.

Projects were funded based on the degree they emulated and reinforced behaviours viewed as morally good. The refugees quickly learned that some behaviours were rewarded, while others were unrewarded and considered problematic. To access the resources in the camps, and to maintain their inclusion in a humanitarian space, they had to emphasize the significance of national boundaries while reflecting broader humanitarian values. Frequently the development projects functioned to inform the refugees that their social structures were largely unacceptable unless they conformed to particular norms and values presented as universal. These values had to appear to be internalized: the refugees had to strive to reform themselves. The Bhutanese refugees actively presented themselves as morally righteous. For it was righteous Bhutanese who could access a resource that was scarce: resettlement. They had become developed enough to deserve resettlement in wealthy, Western countries. This development and restructuring often required a radical break from the past.

A Moral Marriage

This restructuring was particularly blatant regarding marriage customs, where well-meaning interventions yielded unintended consequences. For the Bhutanese refugees, national boundaries have not always been a central point in dictating their personal relationships—particularly marriage. This failure to marry along national lines was a factor that led to their exile: in 1988, the Royal Government of Bhutan stripped the citizenship of any person born to parents who were not both Bhutanese (Citizenship Act 1985; Lee 1998; Saul 2000). The government also gave written notice that non-Bhutanese spouses must evacuate the country within thirty days. While the government maintains these measures were applied to all Bhutanese, the south Bhutanese were the group that frequently married across national lines. These marriage practices were viewed as delegitimizing the nation-state of Bhutan.

In the camps, marriage practices were similarly deemed problematic, becoming a site of humanitarian intervention and evaluation. The UNHCR and IOM, through their liaisons with countries of resettlement, held specific concepts of marriage that refugees must reflect in order to participate in resettlement. Marriages, to be legitimate in terms of resettlement, must conform to three guidelines: they must be between two adults over the age of eighteen (as defined by UNICEF's 1989 Convention on the Rights of the Child); respect national/humanitarian boundaries (Bhutanese–Bhutanese); and be between two people (while the UNHCR and IOM recognize same-sex partnerships, the Bhutanese I worked with understood marriage as occurring between a man and a woman or women).

For the Bhutanese, caste plays a central factor in the selection of a marriage partner: marriage is expected to observe and perpetuate caste lines. For example, a Brahmin is expected to marry a Brahmin. A marriage partner is selected by the senior male in the family unit in consultation with a priest or priests. These priests evaluate the respective astrology charts to check compatibility and select an auspicious marriage date. Historically, marriage occurred at a very young age, around age seven for the bride and nine for the groom, though the bride did not enter her husband's home until she began menstruating. These arrangements were undertaken to ensure the virginity of the bride, in line with patrilineal ideals of descent. Now, due to the changing role of children and conception of childhood, these child marriages technically no longer occur. Virtually every participant strongly disavowed child marriage as backwards or old-fashioned. Even those who were child brides themselves described it as "uneducated" behaviour. Many explicitly quoted the Convention on the Rights of the Child when discussing their views. In order to respect the international norms and values they were taught in the camps, they stated that arranged "promises" rather than "marriages" occurred. In the camp, a similar process of selection and astrological consultation occurred. Again, the priest performed a ritual, but it was not the marriage ceremony. The degree to which this was binding was difficult to discern. Some of the marriages that took place when I was in the camp saw the couple betrothed from childhood. Most claimed the ritual was loosely binding, with space for negotiation once the children came of age or circumstances changed. Though this may appear to be an unquestioned step in the right moral direction, it was still an incursion into the family of a refugee. These were radical transformations occurring in a space that claimed to be apolitical.

Marriages between a refugee and a local were subject to intense scrutiny during the resettlement process. This led most refugees to criticize these arrangements (previously a rather mundane if not ideal configuration) as corrupt or self-serving. The venomous criticism of these arrangements was, however, largely abstract and carried a performative dimension. When participants spoke about specific families resulting from a mixed refugee–local marriage, it was with sympathy. There was genuine concern that families would be left behind or forced to divorce. Divorce, though it happened in the camps, was not widespread and was considered a recent phenomenon. Participants were unable to recall divorce occurring in Bhutan. The only couples I met who divorced did so because of the restrictions on resettlement. Many participants were keenly aware that refugees, particularly with resettlement, had a degree of security that the locals were theoretically not entitled to. It was only through maintaining this image of difference that they could continue to access the basic resources granted to humanitarian subjects. A precarious balancing act was being undertaken as the refugees grappled with pre-existing and introduced value systems.

The impact of the adoption of values that favoured national-based marriage practices and monogamous marriages is best illustrated through two different polygamous families, the first a young Nepali woman married to a Bhutanese refugee in the camps, and the second a long-established polygamous family who married in Bhutan. While in the camps, a man approached me, concerned that his neighbour did not have a hut allocation—generally a clear indication they did not have refugee status. I went to the hut and met a nineteen-year-old girl with an infant in her arms, her daughter. She began to explain her situation. She grew up in a remote village in the neighbouring district before entering into an arranged marriage with the man who used to live in the hut and was the father of her child. He was decades older and an esteemed man in the refugee community. Generally, families worked strategically around the "one wife" requirement, registering second wives as aunts, sisters, or daughters to ensure all are resettled together. For this young wife, her local status would have drawn out the resettlement process considerably. Her husband told her they would get a divorce to expedite the rest of the family's resettlement process but that it would not be a real divorce. He promised to send money to support her in the camp, where her daughter was registered but she was technically trespassing. Months later, the money had still not arrived. Attempts by well-meaning neighbours to contact the husband through his extended family proved fruitless. The young wife had Nepali citizenship, but returning to her natal home with a child but no husband would be exceptionally stigmatizing. At a very basic level, she lacked the means to purchase a bus ticket home, though the neighbours had been considering pooling funds to purchase the 150 rupees ($2 Australian) ticket on her behalf. Here, the tension between competing value systems was acute. As a junior wife, she would have traditionally enjoyed the same (or similar) status as senior wives. A twenty-six-year-old woman explained the balance between multiple wives to me: "In the community, both are regarded as equals because they feel they should be equal. It is not her fault (new wife) that she married the man" (Bhujel female, interview, Nepal, 2013).

The wayward husband is mirroring an internationally supported system of values in which monogamy is considered the morally correct form of marriage. Previously, such renouncing of the marriage had strong social repercussions. The husband would have been heavily stigmatized and forced to either maintain his wife financially or find a suitable new partner for her. Conforming to the standards of international organizations undermined existing social institutions. Though undermining these institutions was viewed as a way to help women, it ultimately further marginalized those it claimed to be helping.

Even polygamous families registered as Bhutanese refugees faced difficulties. One such family included a gregarious husband and two doting wives. The first wife was badly injured in a fall a few years after they married in Bhutan. At this point, the husband and wife had one child. It was very difficult for the wife to care for the household in her semi-handicapped state, and she seemed unlikely to bear more children. Even basic tasks, such as fetching water from their well and cooking, were no longer possible. The husband attempted to help, but agricultural activities absorbed much of his day. They decided a second wife was necessary in order to keep the household functioning. A suitable distant female relative was selected. The three adults, now with several children, lived together in Bhutan, fled the country together, and shared a hut in the refugee camps. When I met them, the patriarch was in his seventies and his wives in their mid-sixties. Their communal children had resettled to the United States, and there were new grandchildren. The parents hoped to join them. In order to do so, the husband had to divorce his second wife to whom he had been married for close to fifty years. The seventy-year-old man explained his experiences:

We wish we could go together. I requested at IOM not to leave any family here and to take the family together. I have been treated well there but I want to go with my other family members, I am sad with this separation for resettlement. Nowadays, after three years, some of the separated families are still here and have not been reunited. We're very worried, that is a long time. I am not sure of the actual number but many people are separated like this, people are really worried they will die here and not meet their family in the new country. It is very important to keep families together, they should not be separated but if separated they must be taken very quickly and resettled in the same house together (Brahmin male, interview, Nepal, 2013).

What emerged from these transformations was an imposed structure of marriage that mirrored the institutions of the governing bodies. Child marriage was no longer practised and marriages respected national divisions. Further, there were few polygamous families under the age of thirty, which reflected the broader ideal of monogamous marriage. Every refugee I spoke to in Nepal (regardless of caste, religion, or age) maintained that polygamous families were still considered equal to monogamous couples; there was no stigma or sense that the practice was morally questionable. However, it appeared to have essentially stopped. Younger couples explained it was too expensive and caused problems with resettlement. While the UNHCR did not directly interfere with the family structure of refugees, the programs promoting human rights in the camps criticized polygamy. A framework of human rights asserts, "Polygamy violates the dignity of women. It is an inadmissible discrimination against women" (Human Rights Committee 2000, para. 24).

The evaluation of marriage practices as appropriate or inappropriate was a moral exercise that underscored the forms that humanitarian governance takes in the refugee camp. The promotion of particular models of family was an intimate intrusion into refugees' lives. These intrusions reflected moral decisions masquerading as administrative policy. Many speculated that adopting these arrangements would make life in new countries easier—crucially aware that if their social norms were deemed morally corrupt, it would legitimize their exclusion. Most participants worried about their reception in new countries, how they would be judged or what values (if any) would be selected as morally legitimate. These concerns were the extension of an experience that had been occurring for the last two decades in the camps as they learned what it meant to be a refugee from Bhutan, in Nepal. Administrative decisions transformed the institution of marriage in the camps. These transformations were done under the premise of protection (either protecting children, protecting women, or ensuring scarce resources for deserving refugees). In doing so, camps reproduced a hierarchy of inclusion and exclusion that mirrors the values of the countries that provide funding for the camps. Refugees found themselves in a tenuous position between the two statuses.

Conclusion

The camps fostered a unique image of the Bhutanese refugees as deserving humanitarian subjects. In doing so, a distinct political space was created within an apolitical, humanitarian area. The Bhutanese, as a refugee community, had not only been created in the camps but also consistently moulded into a specific image. The understandings of the international community's expectations of refugees contributed to the maintenance of the image of a "Bhutanese refugee community." Presenting the correct, coherent image translated into support. This conformity also obscured drastic power imbalances. The models of behaviour and values promoted in the camps often meant that international ideals went unquestioned as morally right.

Camps became a legitimate site of domination—legitimate due to lofty humanitarian ambitions such as helping refugees, supporting the vulnerable, and seeking universal equality. This required that the target of these ambitions transforms their values, social norms, and the most intimate of relationships. It is worth returning to the celebration of Bhutan's National Day that began this chapter. When the morning of the big day arrived, I was in the camp early to enjoy the celebrations. Walking to the designated field with my research assistant, I heard the sounds of families waking and starting their days. Arriving at the field, there was but a handful of children kicking a soccer ball in an ad hoc manner. Perhaps I was too early, or maybe the fog kept people away; it was a dreadfully cold morning. My research assistant was not sure where everyone was. He had not attended a celebration of Bhutan's National Day for quite some time. We went back to his hut, drank tea, chatted, and returned to the field a few hours later. As the day stretched on, it became obvious National Day was not going to be celebrated. The lead-up to the event was largely a performance, an attempt to reflect the specific image of a nation-based community that had been created in the camps. Maintaining this mask of community allowed them to present a concerted and strong political voice based on a refugee identity.

Maintaining an image of conformity required that internal divisions became deliberately obscured. The image of deserving Bhutanese refugees was a political arrangement that required considerable internal management, as the following chapter illustrates. However, though the ideals of the managing institutions permeated the social spaces of the camps, these ideals did not necessarily dominate the daily interactions of the refugees. Rather, the refugees took part in a partial and strategic subordination to further their own political goals.

4
Behind the Performance

THE PREVIOUS CHAPTER argued that humanitarian expectations of refugees mould the Bhutanese. The creation of a morally righteous Bhutanese community facilitated the management of the camps and the camps' inhabitants. However, the Bhutanese arrived in Nepal with existing social structures and moral institutions. As a result, adoption of introduced norms and values was partial—though the appearance of completely internalizing values was viewed as essential to the refugees' ongoing support in the camps. The performance of this righteous identity is best interpreted as an onstage construction, with very different behaviours characterizing backstage activities. The refugees were trying to maintain cultural values that held meaning for them while simultaneously demonstrating their willingness to adopt the values espoused by the managing organizations. It was a tenuous balancing act that raises broader questions regarding recognition and humanitarian value systems.

On my first day visiting the camps, a high-caste Brahmin provided me with a list of close to fifty people that he considered were suitable for me to work with. Over the subsequent weeks, my research assistant led me through this list of predominantly high-caste Hindus who served as community representatives. The people I met with spoke earnestly about Bhutan, the morally repugnant locals, and the importance of serving their community. The stories reinforced each other, emphasizing the shared values between the refugees and the managing institutions. This list proved both helpful and problematic. It illustrated very quickly that the highly orchestrated interactions I was funnelled towards reflected the ethos and behaviours the more powerful refugees felt were essential to maintaining humanitarian support in the camps. The people who appeared on the list were not actors. While these were genuine experiences, they were displayed as representative of the group in its entirety, a performance of Bhutanese "refugee-ness."

The list became a hindrance as I attempted to move beyond the performance, to grasp what social structures were both supported and obscured in these strategic representations. Accessing people who were not on the list was difficult: my desire to speak with lower castes or religious converts was questioned and at times curtailed. Similarly, moving conversations beyond the themes considered acceptable by the more influential refugees proved challenging. For the Bhutanese, attempting to define themselves as deserving humanitarian subjects necessitated elaborate measures of performance management (Goffman 1959; Berreman 1962). This performance simultaneously enunciated divisions that were considered acceptable by international managing structures—particularly cleavages based on national affiliations—while masking internal hierarchies and divisions. The image they present is akin to Adams's (1996, 40) analogy of the two-way mirror: a cultural exchange of values in which the ideal "Bhutanese refugee" is both produced by the organizations running the camps and by those who live in the camps.

In the camps, the divisions between undeserving local and deserving refugee emerged as a morally acceptable division partly because it underscored the legitimacy of national boundaries. In contrast, service providers flagged the caste system as morally reprehensible. Participants came to understand its practise as a threat not only to national constructs of Bhutanese-ness but also to the assumed solidarity of suffering as the moral construct of the refugee. This had the effect of normalizing some hierarchies while problematizing others. To analyze these interrelated processes, this chapter first interrogates the presentation of boundaries between refugees and locals. Next, it examines how the group's internal divisions were managed. This approach demonstrates that humanitarian ideals encouraged particular moral behaviours of refugees that reflect the ideals of institutions funded by the Global

North. Refugees learned that concealing alternative constructions of themselves was a necessary aspect of life in the camps.

The Thin Line between Local and Refugee
Participants consistently juxtaposed the image of a morally righteous Bhutanese refugee against the morally corrupt Nepali citizen. There was an often-repeated equation between the lack of moral righteousness on the part of the locals with the payment of wages by the UNHCR and international NGOs. Refugees criticized the payment of locals as fostering a "thirsty" people: cash created dissatisfaction and an expectation for more. This understanding mirrors the decisions of the managing institutions. Technically, the Bhutanese refugees were not allowed to move—much less work—outside the camps. Reinforcing this ideology was the humanitarian ideal of "community service"; these policies and programs effectively created an image of the ideal Bhutanese refugee that did not strive for economic gain. Hence, the key aspect of the Bhutanese performance in the camps was a desire to serve their community. The locals were paid for work performed, while refugees volunteered. This fostered the appearance of two groups, the locals and the refugees, existing parallel to each other.

For the refugees, maintaining this image promised access to resources and support not available to the local population. Thus, working outside the camps was considered a threat to a righteous Bhutanese identity. The young, educated, and politically minded frequently described "outside" work as demeaning or exploitative. They maintained that interacting and working with locals was a threat to the Bhutanese moral legitimacy: "spoiling" was the key term in circulation. This understanding, though not entirely without justification, enhanced the divisions between local and refugee. It was a simplified understanding of what were complex and subtle social interactions. In turn, this mirrors the way international agencies attempted to order the camps along national boundaries. The reality was much more blurred. The refugees' participation in paid labour challenged the performance of difference and complicated understandings of what it is to be a Bhutanese refugee.

The pothole-riddled road between Damak and the camps wound its way through picturesque paddies. On the rare day that the smog lifted, you could glimpse Kanchenjunga, the third-highest mountain in the world, or Makalu, another peak topping eight thousand metres. The more common sight was fellow commuters: piled high on the roofs of buses, packed snugly into minibuses, methodically powering sturdy single-speed bicycles, or walking. The first bus to make the trip left around 6:00 a.m., covering the journey from the camps into the town of Damak in 10–15 minutes. It was a government bus, resembling a decrepit but brightly painted American school bus—the engine rumbling and belching smoke. Government buses ferried passengers back and forth roughly every half an hour. Minibuses supplemented the timetable, with departures every 8–15 minutes during the morning and late afternoon. The fares were the same, but the minibuses were much more crowded, with smaller passengers expected to clamber on laps or perch on wheel wells. I spent several days conducting surveys regarding public transportation use by the refugees. On any given day, close to three thousand refugees moved between the Beldangi camps and the town. A few worked in local shops or restaurants in Damak. Others worked as teachers in local schools. The majority found employment in construction or as agricultural hands. The morning would see an influx into town, while the evening experienced a reverse flow.

While service without expectation of financial benefit was promoted in the camps, volunteering was not always a valid livelihood strategy. During the autumn of 2012, most refugees undertook the fifteen-rupees (twenty-five Australian cents) daily sojourn to participate in the rice harvest. Rice was a defining part of life both in Nepal and for the Bhutanese. Aside from the ubiquitous "*namaste*" (casual hello), the other common greeting was "*bhaat khaayo?*" (have you eaten rice?). On a practical level, it was the dietary mainstay of both the refugees and local Nepalese. It played a role in virtually every ritual—offered to the gods, given as alms, and bestowed upon friends or family to mark special occasions or rites of passage. The social calendar revolved around the life cycle of rice. The festival season coincided with the end of the growing season. In the low-lying Tarai region, both in Nepal and Bhutan, rice was planted at the beginning of the monsoon season in June. By November, the green stalks had turned golden, signalling the beginning of the autumn harvest. The rice harvest commenced by cutting the ripe grass approximately four inches above the soil. The roughly three-foot-long sections were bundled together and tied with pieces of rice stalks.

These bundles sat in the sun for a few days to dry before being carried on the head, back, or underarm to a central spot for threshing. Threshing the rice against a raised, slatted platform that allows the grains to fall through to the ground was still mostly done by hand. These grains were further sorted on large, flat baskets, which were gently vibrated (almost exclusively by women) to reveal small stones, bits of chaff, and foreign seeds.

Harvest days were long, with refugees often beginning the walk to the neighbouring villages or catching the first bus as the sun made its appearance on the horizon. Men and women of all ages participated. There was a sliding pay scale linked to age and gender. At its most practical level, participating in a rice harvest allowed the refugees to earn a little money. The wage for a male agricultural labourer amounted to roughly twenty-five rupees (forty Australian cents) per day. This is in contrast to the 170 rupees a male or the 125 rupees a female citizen of Nepal expected for similar work in the same region of Nepal (Central Bureau of Statistics 2012, 61). The property owner generally supplied a midday meal. Occasionally, cigarettes or alcohol were exchanged as a form of payment.

Those raised in the camps were very critical of these arrangements. The youth consistently described the work as exploitative and demeaning, something that should be avoided at all costs. However, many still did participate, particularly if the family needed money to participate in festivals or finance a social event. Though the wage was substandard, and working as a hired hand was not ideal, older refugees more willingly participated in the harvest. This was particularly true for men. Many older male participants expressed considerable sadness (that occasionally slipped into anger) at not being able to work the land. A Chhetri man in his late seventies explained a widely held perception: "Many of us still long for our country, our fields—we grew everything there: rice, cardamom, and oranges. Sometimes I wish I could take a gun and get my land back" (interview, Nepal, 2012). Participation allowed them to maintain some of the skills that defined their lives before exile. Many worked, much to the dismay of their children, when money was not crucial to the family's survival. I witnessed one interaction with a young, educated male imploring his elderly (though still robust and able-bodied) mother not to work. Money was not a major issue for the family. An older son was already resettled and sent frequent remittance. She did not explicitly say why she wanted to go, smiling and shrugging when asked directly. The mother went to work the land for reasons more complex than the small wage she could demand, just as her son was frustrated for more reasons than his mother simply being out of the house for the day.

Almost universally, the youth viewed agricultural work as a threat to their position as Bhutanese refugees. The Bhutanese generation that was no longer agrarian consciously presented itself as educated and civically minded. A willing participation in manual labour undermined the image of a modern, developed Bhutanese identity. Focusing on exploitation at the hands of the locals helped maintain their deserving status by emphasizing the lack of legal protection they had within the state of Nepal. This redeemed the older generation's somewhat willing participation in paid labour: their parents were being exploited and, hence, further victimized by the locals. The need to frame their participation in this manner illustrates that economic participation was a potential threat to their morally righteous status. This threat became acute in the high-status and well-paid jobs, particularly teaching outside the camps.

Teaching was a profession esteemed by Bhutanese participants. Though teaching was viewed as a desirable profession, the schools in the camps suffered from a chronic shortage of teachers. This was partly due to the resettlement process: roughly a thousand refugees were leaving for new countries each month. The camps were quickly becoming depopulated, and those with higher educational attainments appeared to be the first to volunteer for resettlement. Aside from refugees leaving to resettle, there were also financial considerations that impacted on the decision to work outside the camp as a teacher. The emphasis in the camps was serving one's community without expectation of monetary gain, though teachers in camp schools received a modest stipend from Caritas. (This small wage might be a reason Caritas was considered the best organization to work for.) Schools outside the camps, particularly private institutions, paid considerably higher wages for teachers with a strong command of the English language. Furthermore, schools such as Montessori held international links. These schools were perceived as having the potential to help refugees find work once resettled. Refugees spent considerable sums to participate in certification programs at the "international" institutions. It was difficult to determine if these expensive courses carried credibility outside of Nepal, or even beyond the specific institution.

While the local schools provided promising opportunities, working outside the camp was complicated. Being Bhutanese increased one's chance of gaining employment due to the reputation of the camp-based schools, but it is also evidenced they were not Nepali citizens. Most refugees who worked outside the camps slipped into a subnational category, focused on caste or ethnic affiliations. A female teacher who moved between working in the camps and in broader Nepal explained, "If they're some 100 kilometres away, I can say that 75 per cent don't say they're refugees; we simply say we are from somewhere else" (Rai female, age 20s, interview, Nepal, 2012). Refugees that had worked outside the camp for several years, some for decades, needed to be physically present to participate in the resettlement process. They also had to demonstrate they were, in fact, refugees, despite a period outside of the camps. While this could be demonstrated by identity documents and the testimony of relatives, they may not be on recent camp rosters. In this situation, demonstrating they are genuine refugees became imperative and tied to not only country of origin but also morality.

Refugees recognized this fluidity was troubling. The physical and moral boundary between who was a local and who was a refugee that was so crucial in the camps became less obvious. As refugee situations become more protracted, with 76 per cent of refugees recognized by the UNHCR now spending roughly twenty years in camps, maintaining clear boundaries became increasingly difficult (UNHCR 2022, 22). This is akin to the Hutu context Malkki (1995a) observed in Tanzania. In a similarly protracted situation, Tanzanians living far from the camps did not describe themselves as refugees. This allowed them to blend into the host community. In turn, Malkki (1995a, 1995b) argues the "naturalness" of the nation-state as a means of defining people was brought into question. Similarly in Nepal, the easy traversing of boundaries threatened to blur the line between deserving Bhutanese refugees and undeserving Nepalese locals. National identity and even citizenship were not always the most significant markers of identity. In other words, the perception of refugees is contextual. While citizenship or nationality is significant, these might not always be the most significant aspect of one's identity. In turn, these porous boundaries, which were presented as impenetrable, raised broader questions regarding contemporary imaginings of refugees.

For the Bhutanese, job security, high status, and independence became possible due to the development projects in the camps that promoted education. Yet they ultimately threatened what are key identifiers of a deserving refugee: poverty, dependency, and the centrality of national boundaries. In this manner, it has strong similarities to Gabiam's (2012) observations in relation to Palestinian refugees. For the Palestinian refugees, improving living conditions was viewed as undermining their ability to make demands based on victimization while simultaneously functioning as a symbol of international support for their cause (Gabiam 2012). The links between suffering and political claims become unsettled if refugees appear capable of supporting themselves. The Bhutanese are eager to participate in development activities that illustrate their ability to fit into an international community. However, some of these become a dangerous endeavour when they develop refugees' capacities to a degree that threatens the attributes that mark them as deserving. This highlights the broader metamorphosis of the refugee away from a group in need of political action towards a compassionate approach geared towards the alleviation of suffering (Gabiam 2012; Ticktin 2011).

For refugees who lived outside the camps for several years, many working as teachers, returning to the experience of being "real" refugees required a new way of representing themselves. The returnees I spoke with described their motivations for returning in terms of the broader theme in the camp: they wanted to serve their community. This motivation was morally righteous, yet it illustrated the conundrum these "successful" rather than "suffering" refugees faced. Technically, refugees living outside the camps are entitled to resettlement. Though somewhat locally integrated, they lacked citizenship and were barred from returning to Bhutan: they were still refugees. However, none of the returnees articulated that they returned to the camps to exercise their right to participate in resettlement. Rather, experiencing material deprivation while serving their community was viewed as a requisite of "refugee-ness." It is likely these representations were influenced by my presence and assumptions I had ties to the institutions that had the power to make decisions regarding their resettlement. This, in turn, illustrated the strategies refugees had to use to balance what were described by international agencies as *rights* with what were experienced as *contingencies* based on the perception of deserving refugees. The camp rhetoric promoted self-sufficiency and community while demanding the recognition of boundaries. In turn, this tension between the right to resettle and the imagined refugee who deserves resettlement reflects the following:

> *the evaluative principles and practices operating in the social world, the debates they arouse, the processes through which they become implemented, the justifications that are given to account for discrepancies observed between what should be and what is actually (Fassin 2008, 334).*

In the camps, not only was the distinction between local and refugee problematic in many circumstances, so too was the performance of the Bhutanese as a united community.

Caste as Deformity, Caste as Conformity
Participants consistently drew a direct line between Hinduism and the caste system. As a means of providing order to the social structure in the camps, however, the reach of caste was pervasive. It dictated the kinds of food people ate and shared with others, whether they could drink a cup of tea with someone, and who was allowed into their home. Caste regulated the choice of marriage partners and friends—whom they greeted in public and whom they avoided. It was omnipresent. Caste and ethnicity often were linked together, leading to generalizations about physical appearance. The Brahmins and Kshatriya (more commonly known as Chhetris) were described as tall, fair, and possessing a pointy nose (Brahmin male, age 56, interview, Nepal, 2013). A low-caste man explained to me these distinctions: "I can tell a Brahmin because he looks like a Brahmin. [Researcher asks what he means—how can he tell?] He walks like a Brahmin [participant pulls back shoulders and raises his chin], and it is based on his complexion" (Gazmere male, age 30s, interview, Nepal, 2012). The way people moved through space: their gait, mannerisms, and posture, all were viewed as

sending crucial clues regarding social status. Names, both surnames and given names, also functioned as identifiers: "I can tell a Brahmin when I ask his surname and automatically I know what caste he is from. People don't lie about their surnames" (Brahmin male, age 70, interview, Nepal, 2013). According to the religious texts (Vedas), Brahmins are associated with the head, while Chhetris are linked to the arms. The middle castes, including those now considered Indigenous groups in Nepal, are stereotyped as looking "Chinese," with a flat nose and tan skin. They are associated with the trunk of the body. The lower castes are generalized as possessing a dark complexion and are linked to the feet. Physical characteristics and bodily associations are directly linked to idealized occupational status: Brahmins are priests; Chhetris are warriors; the middle castes are labourers or tradespeople; and the lowest castes perform work that is considered polluting. Higher castes and high caste was generally how participants referred to the ritually pure castes (both Brahmins and Chhetris). However, Brahmins did not always consider Chhetris to be "high caste," but did consider them higher caste than most others. "Lower castes" and "low caste" was the colloquial way participants referred to ritually impure and untouchable castes.

Even though conversations suggested a degree of coherence, the term "caste" must be approached critically. Michaels (2004, 166) argues it encompasses five criteria: it must be a sufficiently sized group, maintain external boundaries, share common activities, provide a sense of belonging, and provide a system with clear roles. There were at times competing opinions regarding where a caste or ethnic group should sit within the hierarchy, highlighting the complex and contested nature of the caste system. For example, most Chhetris I worked with felt they were the pinnacles of the caste system. Yet Brahmins unanimously placed themselves at the top of the hierarchy and viewed Chhetris as being their subordinates. Burkert (1997, 258) similarly observed that, in Nepal, "assignment to particular categories is subject to competing interpretations, as well as change over time." Further, these are not necessarily "primordial, but brought into being through historical process in which deliberate design may or may not play a part" (Whelpton 1997, 70). Hutt (1997, 118) also cautions, "Caste differentiation also varied considerably from place to place." However, caste or ethnic affiliation was a cornerstone of social relationships in the camps and had strong parallels to the caste and ethnic categories put forth in Nepal's 1854 Muluki Ain. This act codified Hindu caste structure, institutionalizing the hierarchy in Nepal, and justified discrimination against some castes/ethnic groups (Yadav 2016).

Table 1: Nepal Social Hierarchy

Hierarchy	Habitat	Belief/Religion
A. Water Acceptable ("Pure")		
1. Wearers of the Sacred Thread		
"Upper caste" Brahmin (Bahun) and Chhetri	Hills	Hinduism
"Upper caste" (Madhesi)	Tarai	Hinduism
"Upper caste" (Newar)	Kathmandu Valley	Hinduism
2. Matwali Alchohol Drinkers (Non-enslavable)		
Gurung, Magar, Sunuwar, Thakali, Rai, Limbu	Hills	Tribal/Shamanism
Newar	Kathmandu Valley	Buddhism
3. Matawali Alchohol Drinkers (Enslavable)		
Bhote (including Tamang)	Mountains/Hills	Buddhism
Chepang, Gharti, Hayu	Hills	Animism/Hinduism
Kumal, Tharu	Inner Tarai	Animism
B. Water Unacceptable ("Impure")		
4. Touchable		
Dhobi, Kasai, Kusale, Kulu	Kathmandu Valley	Hinduism
Musalman	Tarai	Islam
Mlechha (foreigner)	Europe/USA/etc.	Christianity, etc.
5. Untouchable		
Badi, Damai, Gaine, Kadara, Kami, Sarki	Hills	Hinduism

Source: Adapted from Muluki Ain (1854), Whelpton (1997, 2005), Michaels (2004), and refugee interviews (2012–2014).

The modern role of caste was frequently used to justify why some groups held considerable sway in the camps, maintaining positions of power, interacting with international staff, and enjoying a higher standard of living, while other groups struggled. In other words, attempts to maintain boundaries between groups and preserve internal cohesion led to distinct political and social consequences (Whelpton 1997). For instance, Brahmins were rewarded handsomely for performing rituals. The reading of sacred texts was necessary for weddings and funerals. Entire epics must be read an auspicious number of times. The priests who did these readings (sometimes upwards of six times for a single ceremony) received money and other resources as repayment. It was not only an esteemed task but a lucrative one. I was told by participants that a few training centres in India accepted lower castes, however it was not clear how many of the lower-caste refugees had undertaken training. I heard rumours of one such priest who provided services to lower-caste refugee families, but he had already resettled when I was conducting fieldwork. It was clear, though, that his training to be a priest did not change his caste and its associated polluting properties. Higher castes would not consider employing a lower-caste priest. The perception of physical differences was used to justify their lack of employment. It was explained as "to read the Vedas though is very difficult so many people ignore the pronunciations. Especially from the lower castes, they can't pronounce correctly, they have thick tongues" (Brahmin male, age 34, interview, Nepal, 2013). Brahmins maintained that the tongues of the lower castes lacked the flexibility to speak Sanskrit elegantly. In this way, stereotyping became a means of legitimizing inequality. Rather than access to opportunities being perceived as limited due to drastic power imbalances embedded in broader social institutions, these limitations became justified through the perception of physical differences.

Caste was a fundamental way in which the Bhutanese understood the world and their associated place within it. There was great reluctance to accept interference in this arena, hence the steadfast and diverse resistance to discussing the intricacies of the system. The many campaigns in the camps to abolish the caste system have, in effect, made the issue a sensitive one and off limits to outsiders. At the head of this resistance (though by no means the only group opposed to outside intervention in the caste system) was the priestly caste—generally both male and female Brahmins. However, many Chhetris were similarly resistant to change: "The caste system helps keep our community organized and it is good to have the caste system. We're satisfied with our caste and others should be satisfied with theirs" (Chhetri female, age 34, interview,

Nepal, 2012). Their attempts to present an image of a caste system that has disappeared or been radically refashioned relates to their traditional role as the protectors of Hinduism. For most Brahmins, caste and Hinduism could not be disentangled.

The social dominance of Brahmins at times was inadvertently enhanced by ideals promoted by international agencies, namely the emphasis on education and the promotion of community cohesion. Further, campaigns highlighted the importance of cleanliness, and the way interacting with locals can "spoil" Bhutanese identity inadvertently fit within the existing framework of purity and pollution. These were central to the Brahmin castes. As the previous chapter illustrated, in Bhutan, Brahmin males did enjoy greater access to education than any other group. Virtually everyone I spoke with expected Brahmin men to be educated. In the camp, though all castes and both sexes received the same number of years of schooling, the perception was that Brahmins were inherently more intelligent. The notion persisted that people were born with clearly defined roles (in their castes), and these roles demanded particular attributes. A key aspect of the sacred text *Bhagavad Gita*, according to participants, hinged on the various attributes of people within the broader cosmos (Brahmin male, age 56, interview, Nepal, 2012). Vishnu Brahmins are expected to behave in a fashion that reflects the sacred texts: they lead by example and inspire conformity. Brahmins, and particularly the highest-ranking Vishnu Brahmins, are the protectors of the Hindu cosmic order. From the perspective of ritual cleanliness, Brahmins (and specifically Brahmin males) sit at the pinnacle: they are able to interact with their gods much more closely than any other group. Their ritual purity is maintained through bathing, for water is exceptionally important in Hindu cosmology. It is a conduit that ritually cleanses the higher castes and allows them to closely worship the gods.

In the camp, bathing every day was essential for the highest castes. The Lutheran World Federation maintained the communal wells that provided clean drinking water for the refugees. These flowed twice a day: first in the morning and then in the afternoon. Generally women and young girls are responsible for collecting and transporting the water. To avoid the long waits that queuing for water often entailed, the women would line up their containers in anticipation of the water collection. These containers "held" their spots so they could continue with their daily activities: preparing food for the family, tending the garden, sweeping their homes, or perhaps beginning to soak the laundry if there was water remaining from the previous day. While it was essential that women did not lose valuable time waiting with their containers at the communal wells, this also meant that containers left behind risked contamination due to their proximity to ritually polluting containers. Thus, the communal water pumps were a place where the contradictions between the assertion that caste is no longer pertinent and the desire to maintain caste-based social hierarchy became obvious.

Several higher castes complained that lower castes were deliberately attempting to touch them when they went to collect water, consequently polluting not only the person collecting the water but also causing the water to be "impure." Participants recognized the Lutheran World Federation would not consider these complaints. A few participants maintained they did request separate wells but were informed their motivations were not appropriate. In order to overcome this situation without sullying the image of a united community, alternative wells were developed. These wells were deliberately hidden from international staff (and researchers) working in the camps. I was working in the camps for several months before I noticed a hand pump beside a Brahmin's home. The pump was on a fenced, elevated platform, obscured by plants and further protected by a large padlock. The family, whom I worked with on a daily basis, became elusive when I inquired about the well. Initially, they avoided my questions but later asserted they had to protect the well because animals could damage it. While there was some animal husbandry occurring in the camps, the damage a straggly chicken or a waddling duck could do seemed disproportional to the fortifications. Eventually, it became evident this well allowed the family to access a source of water "unpolluted" by the lower castes without appearing discriminatory.

The criticisms of the caste system by international organizations in the camps functioned to further regulate the group and potentially legitimize further interventions in their lives. Hence, when lower castes deliberately tried to pollute the wells (from a ritualized standpoint), this was perceived as not only corrupting the Hindu social world but also undermining the image of a united community by drawing attention to a still functioning caste system. The lower castes were acutely aware that the higher castes did not have recourse for such complaints through the governing institutions. A Brahmin woman in her thirties explained the strategy she employed to practise the traditional system while abiding by the camp rules:

> Before, a Brahmin could say, "You're no good. Don't touch me, don't come to my house." Now we have to be quiet but they [the lower castes] understand. They come to my house and I will give them some tea, no problem, but they know they need to stay outside. We don't say no but people understand. Now we'll get in trouble if we say, but everyone understands. They won't try to enter my house. After they leave, I'll wash the cup extra times: special washing. Many people won't let them sit in a chair but I will let them. I would wash this chair many times though: special wash (interview, Nepal, 2012).

Participants' attempts to continue their cultural norms while appearing to conform to international ideals should not be dismissed as hypocritical. The higher castes were attempting to merge the social institution that very much defined their social relationships with the "rules" of the international agencies in the camps. This underscores the persistent imbalance of power within the group but also in relation to humanitarian agencies managing the camps.

The higher castes were quick to publicly assert that the caste system was outdated, unpractised, or a product of the corrupting influence of the local population. One high-caste Brahmin family I interacted with almost every day in the camps—including acting as the "official" photographer for their daughter's wedding—helped me develop a survey for use in the camps. As we worked through the questions regarding caste, the family consistently asked if they were getting the answers right. They were acutely aware that caste was flagged as a social problem by international agencies. The dominant rhetoric by the higher castes was that the system is gone. Their liminal status as refugees required them to publicly disavow a central way of understanding their world. This requirement demonstrated that attempts by humanitarian organizations to create a more perfect world morphed into a form of governance that not only cares for but controls its subjects (Barnett 2011). This more perfect world included "injecting values that are presumed to do a better job of improving well-being" than existing social structures (Barnett 2011, 231).

Mixed Marriages

Conversations regarding extra-caste marriages, deemed "mixed marriages" by participants, emerged as a means to discuss the contemporary impact of the caste system. The following interaction with a high-caste Magar male in his twenties illustrated the gap between rhetoric and reality:

> *Nowadays, people are educated. It is loose compared to earlier days in Bhutan. People now that have no concept of caste, not thinking from this caste or that caste but all equal.*
>
> *[Researcher: If you married outside your caste, how would people treat you?]*
>
> *People would deny to visit my house and wouldn't eat anything that I touch. We would get bad treatment as well (interview, Nepal, 2013).*

The idea that the group had been educated to have "no concept of caste" was an attempt to show they had internalized the values of the humanitarian groups in the camps. They were one community where everyone was equal: their emancipation from problematic social structures was complete. This veneer of community began to break down when the specifics of caste dynamics were discussed. As I moved beyond the high-caste gatekeepers, it became evident the image of a united community was part illusion, part strategy, and part aspiration. For the higher castes, and virtually every Hindu participant, religion, caste, and community were explicitly tied: if they changed religions, they ran the risk of losing their community.

If a higher-caste Bhutanese married a lower caste, irrespective of their gender, their former caste status was lost and they were no longer ritually pure. Such a couple's children would inherit lower caste status, and they would be barred from entering higher castes' homes. This represented a significant rupture with social norms, as the wife was expected to join the husband's family and provide a significant domestic contribution. Mixed marriages could result in high-caste families performing burial rights for the wayward child and ceasing all forms of communication. This was a devastating and hurtful situation, both for the children and the parents. Sometimes this was a permanent arrangement, while in other families the birth of a grandchild was enough to rekindle a relationship. This new relationship was considered a positive development, but the lower-caste member was still treated as impure and, as such, was forbidden from eating with the family. The treatment of the child varied by family, but some higher-caste families made exceptions to the strict dining rules generally observed. For most who decided to marry outside their caste, both the lower- and the higher-caste families rejected them. When discussing mixed marriages with me, participants invariably spoke of a high-caste woman who married a low-caste man while in Bhutan. This couple was thoroughly rejected by the broader group. They had to build a home away from their family and communication essentially ceased. Virtually every participant in the camps knew about this transgression. Because this individual incident was so notorious, and participants were unable to recall any other instances, it suggests that in Bhutan mixed-caste marriages were rare. However, in the camps, there had been several (perhaps even a hundred) of these unions.

"Love marriages" were frequently attributed to the schooling system in the camps. The Caritas-run schools had exceptional participation rates for males and females across all castes. The schools in the camps created an environment described as "everyone mixed." Groups that previously had little direct interaction found themselves studying together. When I inquired where couples who eloped in a love marriage met, they almost invariably pointed to the school. Love marriages were considered problematic, but this varied based on the degree of difference perceived between social, caste, or ethnic unit. For example,

I did the love marriage and community felt this not good to do. I am Gurung [middle caste] and husband is Sarki [untouchable]. We met at school, at first my parents scolded for marrying out of caste but they were Christian and suggested that we become Christians otherwise in future we would be out of the scene, not be able to make a good future so my husband converted—I was born Christian. We celebrate all things with our church community and we invite our neighbours as well (Gurung female, age 21, interview, Nepal, 2013).

In this instance, the Gurung family was reluctant to accept the marriage but provided a pathway for the couple to move forward. There are fewer pathways forward for marriages occurring between those at polar ends of the social hierarchy:

It was a love marriage, inter-caste. I am from Kafle caste [Brahmin], and my wife is Biswa [low/untouchable]. At the time, some people were tricky when we married, we had a big ceremony, but my family did not come. My father and mother, it was quite difficult at first. I had to teach them, the society is not like the previous one and now [sixteen years later] they have changed (Brahmin male, age 31, interview, Nepal, 2013).

While this participant flagged that there had been improvements over the past sixteen years, his wife is still barred from entering his parents' home. When asked what happens when different castes marry, responses tended to be "the community will hate us" (Gazmere female, age 19, interview, Nepal, 2012), or "it is better not to do these things because it will make things very difficult" (Chhetri male, age 30s, interview, Nepal, 2013). Yet these mixed marriages were occurring at rates that at least give the impression of greater frequency than before fleeing Bhutan. In the camps, the possibility of alternative communities not based on the Hindu social hierarchy has emerged. The Brahmin man quoted above explained to me that, after his family performed funeral rites due to his decision to elope with a lower caste woman, "we were alone and someone said maybe we should think about becoming Christian" (Brahmin male, age 30s, interview, Nepal, 2013).

Converting to Find a Community

Christianity was (until 2008) banned in Bhutan (US Department of State 2008). One man who ran a secret church out of his home in Bhutan still used his hut in the camps to hold services. This was the only Christian I met who practised *before* exile. Since then, a striking number of camp inhabitants had converted to Christianity. The Beldangi camps (I, II, and Extension) had a population of approximately twenty-six thousand during fieldwork. My visits to twenty-nine churches revealed that roughly three thousand people actively attended Christian services. The minister or other church official generally provided the number of members. I was concerned that there would be an inclination to inflate numbers because of the presumption I was affiliated with Christian churches due to my ethnic/national background. To overcome this, I also examined their rosters when available. I was told there were fifty churches in the Beldangi camps, but either I was not able to visit all of them or they did not have rosters available. The actual number of converts was possibly higher. Even at approximately 11 per cent, the number of Christians in the camps was considerably higher than the 1.42 per cent in Nepal (Central Bureau of Statistics 2014, 227) and the 1–2 per cent estimated for Bhutan (US Department of State 2008, 2). There were few Brahmins converting, and those who did convert tended to do so because of a mixed marriage. The middle castes represented a sizeable number of converts, though a much greater number of this group were converting to Buddhism or Kirati (as discussed in the following section). The lower castes, particularly the youth and women, were converting in very high numbers. A lower-caste participant spoke candidly regarding his motivation and experiences after converting:

I changed to Christianity because there is equality in Christianity; there isn't a caste system. I wasn't interested in the previous religion (Hinduism). People have their wedding at the church and when we have a birthday party, people from our church come to our house to celebrate. We have a very strong community…I found that in that church, any issues between people will get support from all castes. Before, if a Biswa [untouchable] died, the community wouldn't come to support the family, but in the church everyone will (Biswa/Christian male, age 24, interview, Nepal, 2013).

Another convert articulated similar thoughts:

I was Hindu in Bhutan but changed to Christianity. There were many rituals in Hinduism, and it is better with Christianity. I feel there is more equality, there will be a community that treats everyone equal, that is

> why I decided to join. I am happy with the new community. We celebrate most things with our church community—either in our home or at the church (Kami/Christian male, age 47, interview, Nepal, 2013).

The reasons for changing religion varied, and the perceptions of those converting were similarly complex. Yet the promise of equality and gaining access to a community that would not treat them differently due to their caste was repeated as the primary motivation. Some members of the lower castes, particularly the elderly, were not eager to change religions—even if their children were. One lower-caste family saw all their children convert to Christianity, but the parents remained reluctant. They voiced the same concerns as the higher-caste Brahmins: that converting would lead to a loss of culture and jeopardize their future beyond this physical life. None of the higher-caste people I spoke with entertained the possibility that people were converting because of discrimination. This understanding undermined the camp ideals they strove to reflect: that they were a united community. Rather, higher castes explained it as "the will of the individual" (Brahmin male, age 24, interview, Nepal, 2012), or "they think they will get more benefits when they resettle" (Chhetri female, age 30s, interview, Nepal, 2012). There were certainly people who felt conversion had strategic value, but this should not be understood as manipulative. Rather, several converts described their decision as an act of reverence to the nation that was resettling them. Having experienced thirty years of exile for what many viewed as an attempt to protect their religious beliefs, they did not want to risk another expulsion. Further, they consistently learned through the camp-based programs the importance of national identity. In the minds of virtually everyone I spoke to, they viewed the United States, Australia, and Europe as primarily Christian countries. However, the high-caste position that people were converting only to enjoy preferential treatment was a clear attempt to undermine the moral legitimacy of converts. Converting to escape persecution or discrimination carried very different connotations than converting to access additional resources.

Higher-caste participants claimed the caste system maintained social order and group cohesion, as caste provided architectural integrity to the group. Part of this understanding linked to the way the higher castes conceptualized the role of the lower castes. High-ranking priests explained that the low castes are valued members of the Hindu cosmic universe as the vessels for inevitable impurities. Of course, those who are pure cannot interact with these vessels, but, the priests claimed, that did not diminish their importance. Participants implied that people attempting to lose their culture via conversions would regret these choices later. This further articulated the perception of the higher castes that Hinduism was intrinsically linked to an exiled Bhutanese identity. Many Hindus (but particularly the highest castes) viewed lower castes that changed their religion to Christianity as dissenters, destroying the fabric of what it meant to be Bhutanese.

The key holidays in the camps centred on Hindu epics. These holidays affirm social bonds and the centrality of Hindu social values. While some Christians in the camps still participated in these rituals, most viewed them as false celebrations. Instead, they promoted the celebration of Christian holidays, Easter and Christmas in particular. In the weeks before Christmas, the Christians in the camps invited everyone living in the camps to celebrate with them. The use of a written invitation—generally in English—was a distinct break from the verbal invitations generally issued in the camps. The written invitation included a sampling of the Christmas menu, which included beef.

Cows hold a central place in Hindu cosmology—dung is used to ritually purify homes, and milk sustains families as a primary source of protein. Further, oxen were crucially important in agriculture for preparing rice paddies and later converting paddies to buckwheat or millet fields. Historically, abstaining from beef may have been a way for Hindus to differentiate from invading groups (Harris 1966; Robbins 1999). For the refugees, consuming beef was a reaction against the Hindu religious system: converts were participating in a larger process of questioning Hindu norms and values. Further, the flaunting of beef consumption was a display of difference from the nation of Nepal. The government of Nepal forbids the killing of cows (Malagodi 2021). Ogura (2007) and Shneiderman (2010) have similarly noted that Indigenous groups in Nepal consume beef as a symbolic and material means to assert a social position independent of the still strongly Hindu government. Christmas festivities became a resource and a ritual that "can speak clearly and centrally to aspirations towards cultural change and even assimilation" (Baumann 1992, 109). Advertising their consumption of beef was a political stance. It evidenced an alternative identity, different from both the local Nepalese and the most powerful members of the refugee community. The Bhutanese were actively working out competing notions regarding what was considered good and evil, pure and impure (Douglas 2003). It highlighted their perceived difference from an agriculturally based, Hindu population and their aspirations to be actors within an international value system. But this was, again, a complicated association.

Hindu participants maintained that converting to Christianity could not change one's real caste. Many expressed genuine concern that associating with castes that had been considered "for thousands of years" as polluting was not wise. As a female Brahmin explained, "Maybe Christian families will allow but I won't allow low caste into my home, I won't eat with the lower castes" (age 30s, interview, Nepal, 2012). For the converts, the change promised a degree of social acceptance previously denied to them. The Christian churches emerged as a means of partially shedding one's prescribed role and its associations. In this respect, it was a social and political stance. What was occurring was more complex than adopting symbols of the West. These converts were embracing and creating a new Bhutanese culture. Blatantly consuming beef and disavowing the caste system disputed their socio-political position within a Hindu-based camp hierarchy. This expression of Bhutanese-ness, in turn, fitted more directly with the nation-based understanding promoted by the UNHCR, rather than the religious-based approach employed by camp leaders. Converting to Christianity thus created an alternative community different from either the Bhutanese identity founded on Buddhism that the government of Bhutan promotes, or an ethnic Nepalese identity that has strong Hindu connotations. These changes suggested that, through this process of conversion, people began to understand and orientate themselves in a fundamentally different manner. Camps created new conditions of Bhutanese "refugee-ness."

Rediscovering Indigenous Identities
Once the resettlement process is complete, roughly ten thousand Bhutanese will remain in refugee camps. These are people who have either decided on personal grounds not to resettle or cannot resettle due

to an inability to pass a security clearance. These people (and some already living outside the camps) had begun to take steps to locally integrate independent of the UNHCR and against the wishes of the government. This had led to a strong resurgence in Indigenous identities. One of the participants, a Gurung man, followed Hinduism in Bhutan and spoke only Nepali. Living in the camps had led him to rediscover his Indigenous roots. The Gurung aspects of his identity had increased in importance. Over the years, he procured books discussing Gurung social norms and took steps to learn his "real" language. He also learned that the Gurungs in Nepal are generally Buddhist, not Hindu. He (along with his family) made the conversion to Buddhism in order to move closer to their genuine Indigenous identity.

This man's decisions were not an isolated case. Interviews and observations indicated a high number of non-Hindu religions in the camps. The vast majority of Gurungs, Tamangs, and Sherpas had converted "back" to Buddhism. Similarly, Rais and Limbus were converting to Kirati, an animistic religion. Surveys I conducted in Beldangi II revealed that close to 100 per cent of Gurungs, Limbus, Magars, Rais, Tamangs, and Sherpa were no longer Hindu. Rather, these groups were (re)discovering "locally grown" religions: Kirati is associated with eastern Nepal, and Gautama Buddha was born in Lumbini, Nepal. During the Rana dynasty in Nepal, these religions were considered branches of Hinduism rather than separate religions (Gellner 2007; Whelpton et al. 2008). For the higher-caste Hindu participants, this understanding was still credible. One Brahmin man in his fifties explained that these religions were "almost the same as Hinduism" (interview, Nepal, 2013). For the Bhutanese who converted, these religious alliances were political, self-conscious, and strategic. These "more authentic" identities (Brown 2010) have found new life in the camps.

Though participants discussed changing religions, it was also possible that presenting themselves as largely Hindu was advantageous during the early years of exile in Nepal. It might have been a strategy to emphasize their affinity to a host country that was, at that point, a Hindu kingdom. While the high-caste community leaders still described the refugees as Hindu, smaller subgroups that previously had been subsumed under the label Bhutanese have moved to the fore. From a demographic standpoint, refugees who tapped into an alternative identity discourse are now the majority. For participants not planning on resettling, Indigeneity presented an opportunity to increase their chances of integrating locally. Particularly for older refugees, being able to remain in Nepal after the resettlement process concluded reflected a desire to die in a country they knew, where they felt certain the correct rituals could be performed. Ideally, this was Bhutan, but many accepted that Nepal was a more likely scenario.

Demands based on ethnicity have increased in legitimacy both internationally and locally during the refugees' time in the camps, and "expressions of indigenous identity have been linked to broader social movements" (Muehlmann 2009, 476). In 2002, Nepal formally recognized fifty-nine Indigenous groups under the National Foundation for Development of Indigenous Nationalities Act. Having an Indigenous claim to Nepal may provide political leverage for refugees who want to settle locally. Further, the UNHCR actively promoted the recognition of Indigenous groups as laid out in the 2007 United Nations Declaration on the Rights of Indigenous Peoples. In terms of accessing resources in the camps, an Indigenous angle may have translated into support or funding that previously was channelled through Brahmin-dominated community leaders. Compared to converting to Christianity, adopting these Indigenous religions was considered less of an attack on the fabric of Bhutanese refugee society. However, the concern persisted—particularly in conversations with Brahmin and Chhetri participants—that fractures in the constructed community of Bhutanese may diminish the degree of autonomy they have worked for in the camps. Higher-caste participants expressed concern: "I feel it shouldn't be changed, there will be disturbances if people change" (Brahmin male, age 56, interview, Beldangi, 2012). As Murray Li (2000) observes in Indonesia, these attempts to funnel resources along Indigenous channels represented an opportunity but were not a guarantee. Again, as with broader refugee status, there was a little-discussed process of evaluation and judgment that accentuated some boundaries while minimizing others to create a hierarchy of deserving. The Bhutanese leaders actively cultivated a nation-based identity to maintain access to the scant resources available to refugees. This image of Bhutanese-ness was largely regulated and maintained by the more powerful community members who sought to minimize the Indigenous associations that might undermine their mask of unity.

While both the highest castes and the lowest castes seemed to see moving away from Hinduism as a major decision, the middle castes approached religious affiliation with greater fluidity. There was one group in the camps that appeared to be less transformed than others: the Sharchopas from Eastern Bhutan. This historically Buddhist group had not converted to Hinduism, Kirati, or to Christianity. It had little engagement in the numerous projects run by international agencies in the camp and existed largely independent from broader trends in the camp. Physically, their huts sat on the fringe—apart from the mathematic precision that defined the rest of the Beldangi camps. Perhaps most strikingly different was their steadfast belief that they would be repatriated to Bhutan. They had little interest in working with researchers and worried that speaking with me would compromise their future return. Though I *did* speak with several Sharchopas, the only close relationship I was able to foster was with a woman in her forties who decided to resettle. After living alone in the camp for two decades and receiving no information about the fate of her imprisoned husband in Bhutan, she concluded he had died in custody. This spurred her to pursue a future in a new country. Her decision to resettle set her apart from her neighbours by opening her to a new social landscape within which to orientate. Her willingness to work with an international researcher suggested a fundamental realignment. For better or for worse, she was taking the first steps towards becoming a humanitarian subject—joining the broader community of Bhutanese refugees.

Resettling
During 2012 and 2013, IOM buses full of Bhutanese destined for new countries left the camps in the early morning hours. The departees' foreheads were heavy with *tikkas* made of red powder and rice that well-wishers had bestowed upon them. Their departure understandably evoked sorrow, for there was still uncertainty if or when families would be reunited. Friendships had not been considered in the resettlement process, and the loneliness for some, even if their family was intact, was acute. Participants told me teen suicides were not uncommon. By February 2013, the resettlement process had been occurring for seven years, at a rate of nearly nine hundred per month. By the early months of 2013, already eighty thousand Bhutanese were resettled. An additional

twenty thousand Bhutanese were somewhere in the process. Where people sat in the particular time frames of this process was a common point of conversation and reference among refugees. Remarks such as "how is your process?" or "there is some kind of trouble with my process" punctuated everyday interactions. The process participants frequently referred to was lengthy, with changes to family situations (births, marriages, and deaths) further drawing it out. To begin the process, refugees had to submit an expression of interest to the UNHCR. After expressing interest, an application containing all the names of family members hoping to resettle together was submitted. Once this was accepted, the family would attend an interview conducted by local staff. This interview was part informational and part initial screening. After these preliminary steps, the UNHCR referred potentially eligible refugees to the IOM, which began the processes of security clearances and health screens that would funnel refugees to various new countries. It was during this process of screening and channelling—before the cultural orientation process officially started—that refugees began to form opinions of various countries' expectations of refugees. These opinions also formed through conversations with members of their immediate family, extended family, and friends that already were resettled. Most, if not everyone, had contact with people already resettled, and some of the first refugees to resettle in the United States had returned to the camps for visits or marriage. This both created and maintained communication links between the camps and countries of resettlement.

Generally speaking, participants viewed the United States as desiring refugees who were able to become economically independent very quickly due to a lack of social safety nets provided by the government. Australia, on the other hand, was viewed by participants as deliberately seeking refugees who needed additional social support, either due to poor health, advanced age, or life experiences. The extra support afforded by Australia's welfare system was weighed against the perception that finding work in Australia would be extremely difficult compared to the US. While refugees also had opinions of New Zealand, Canada, and various European countries, due to my personal links with the US and Australia, these countries were most frequently compared. Refugees were allowed to state a preference regarding their country of resettlement, and existing family links were given priority. Some refugees were able to present themselves strategically and build a case for a specific country by emphasizing health concerns or a particular skill set. These refugees attempted to promote a particular image relevant to their resettlement needs. How much these efforts actually bore on resettlement decisions was not clear—my impression was it was an attempt to feel like they had control over a process in which they had very little say. Ultimately, the IOM, in conjunction with the respective national governments, decided the destination of refugees. Yet the impression that refugees might have some input in the decisions encouraged another form of performance.

Once a refugee was partnered with a country, a date was selected for departure. This date was occasionally known a few months in advance, but generally the space between receiving the date and departure was brief: weeks at most. During this period, a three-to-five-day cultural orientation took place; shopping for goods to accompany the move occurred; and farewells were made. Some refugees attempted to make pilgrimages to particular holy sites in Nepal, while others arranged to see family living in Nepal, India, or Bhutan before departing. While a few did make illegal visits to Bhutan, it was more common for family still in Bhutan to covertly visit the camps. Once the morning of departure arrived, the bus was boarded with much fanfare and emotional outpour. The bus would take them to nearby Biratnagar, where they boarded a charter flight to Kathmandu. The refugees would spend two to four nights in Kathmandu, staying in IOM facilities as they received their international travel documents. Here, they would also learn what to expect on a long-haul international flight, undergo a fitness for travel evaluation, and have a final cultural orientation regarding their new countries.

This long, and at times frustrating, process marks the Bhutanese as the elite of global refugees. They have been consistently praised by the UNHCR and international NGOs. Participants largely understood resettlement as a reward for their good behaviour. Participants were well aware of their status and the precariousness of that position: there are literally millions of other refugees around the world hoping to leave camps and start anew. At stake in the performance of being deserving humanitarian subjects was their future. In turn, this reflected global power disparities—geopolitical sentiments that rewarded some while denying others. This process ultimately affirmed the altruism of the wealthy countries while critiquing the social structures of the less affluent.

Conclusion

Refugee status is a legal right, but becoming a deserving humanitarian subject is largely contingent on particular behaviours. It is also a status that can be sought after and competed over. It becomes an overarching tension as participants attempt to inhabit the role of idealized refugees when they are a very real human group with internal struggles, contradictions, and shortcomings. Camps reproduced hierarchies of inclusion and exclusion: refugees could be included while locals were excluded. However, visions of unity are misleading. Refugees must balance the competing hierarchies of the nation left, the nation they are housed in, and, for the Bhutanese, the nation they will be joining. The nation emerges as a moral category that forms and transforms its subjects, past and present. However, additional hierarchies also demand interrogation. In particular, the ways humanitarian and pre-existing norms and values converged in the camps to produce new understandings of the world. Examining the interaction of these various social constructs illuminates the experience of life under humanitarian governance. Life in the camps becomes a site where competing notions of the world—how community and difference are understood, celebrated, or obfuscated—are negotiated. These tensions make

> *explicit and intelligible the evaluative principles and practices in different societies and contexts, of analysing and interpreting the way social agents form, justify and apply their judgements on good and evil (Fassin and Stoczkowski 2008, 331).*

The Bhutanese deftly negotiated expectations while simultaneously curtailing their reforming efforts. The very specific examples analyzed here are indicative of a broader process of judging, condemning, and intervening—a process that is obscured by the morality of humanitarian values. The evaluation of people as deserving or undeserving is a moralizing discourse. To maintain a morally righteous status, the Bhutanese must appear to reflect the norms and values deemed acceptable. A successful performance obscures complex forms of social organization. As the refugees boarded the plane in Kathmandu bound for Australia, some of the lessons learned in the camps receded in importance as new expectations emerged. Politically, refugees were called upon as a means of evidencing Australia's righteousness. As we shall see in subsequent chapters, this public morality was necessary to manage the boundaries of the country and people within them.

5
On the Threshold of Australia

DECADES OF LIVING IN THE CAMPS taught the Bhutanese that reforming cultural norms and institutions to mirror international values was part of the refugee experience. They hoped that transforming into the right kind of refugee would ultimately mean they could stop being refugees. In Australia, the Bhutanese found themselves in a different kind of political limbo, with new power relations to negotiate and distinct expectations to navigate. In February 2013, these expectations were clearly conveyed to the Bhutanese when an Ethnic Leaders Forum in Adelaide hosted the assistant director of the Department of Immigration and Border Protection. The presentation's purpose was to explain recent changes to Australia's onshore visa processing program. The speaker informed the group that the department had decided to prioritize refugees coming from UNHCR-run camps by expanding their offshore Humanitarian Visa Program. One male refugee at the forum asserted that the offshore process through the UNHCR took a very, very long time. The assistant director explained this was due to the large number of people attempting to come to Australia independently as asylum seekers:

> *The reason why, what is linked back to this issue: onshore protection. Reason that offshore has slowed down is because of high on-shore visas. The UNHCR has the highest numbers ever seen, but processing time*

has increased because our numbers are capped. We're hoping now the processing will be faster (Ethnic Leaders Forum, Adelaide, 2013).

He explained further that, for people arriving as asylum seekers rather than UNHCR-referred refugees, applications lodged to reunite their families would be given the lowest priority. On the other hand, those who came through the UNHCR system attempting to reunite with family still in camps would see their applications moved to the top of the queue. The service providers present were informed they should not be assisting asylum seekers who had not been verified, and that resources should be going to refugees.

This presentation created a clear dichotomy between refugees who arrived using what was described at this forum as "the appropriate" avenues and those who attempted to make their own way. The former deserved the refugee label and associated social benefits, while the latter were subject to suspicion as undeserving of benefits. Beyond the factual aspects regarding changes to the visa program, the participants also picked up on more subtle lessons. Namely, attempts to act as agents of their own destiny (during this presentation, it was framed as deciding to get on a boat) would not be rewarded: waiting in a camp was the *only* acceptable way to legitimately gain admission to Australia as a refugee. The message was also that only those coming through the camp system have the appropriate backstory—morally, they were closer to Australians due to their apparent respect for national boundaries. Their moral superiority translated into support and, possibly, acceptance.

This particular forum was not the only space where Bhutanese learned that a hierarchy existed among new arrivals. Between 2012 and 2014, Australia's media was saturated with images of asylum seekers arriving on boats and being detained in various centres while their cases underwent evaluation. These media representations, paired with the rhetoric of "queue-jumper" juxtaposed with the genuine refugee, influenced the way participants came to understand themselves in Australia. Narratives like these made asylum seekers seem less human than Australian citizens (Bennett et al. 2017). The Bhutanese interpreted these as clear signals that if they wanted to see their families reunited, they needed to promote themselves as ideal refugees. They hoped that by presenting themselves as the most deserving and best behaved, it would sway the government to accept more Bhutanese from the camps. The strategic portrayal of "refugee-ness" to service providers and the broader Australian population was an attempt to influence migration policy. To do so, the Bhutanese capitalized on portrayals of the passive refugee: having suffered in flight, languished in camps, and as gracious recipients of Australia's generosity.

Examining the distinct situation in Australia illuminated broader practices in the global moral climate. Compassion justified the admittance of refugees and in the process obfuscated the consideration that asylum seekers had a legal right to seek refuge. This chapter argues that accepting refugees allowed Australia to position itself as a charitable nation. Refugees were cast in a new role largely in opposition to politically or popularly salient stereotypes of asylum seekers. The exclusion of asylum seekers, who were depicted as greedy or socially deviant, became justified by a need to protect citizens and provide for those deemed deserving of Australia's support. This charitable positioning, in turn, constructed refugee societies in a particular fashion that assumed a lack of social or political institutions. This minimized the strengths and capabilities of refugees, effectively relegating them to the periphery of Australian society.

The Politics of Regulation

Who is admitted into Australia and how admission is regulated have a long, ever-changing history that has persistently marginalized the Aboriginal population. However, an in-depth discussion of settler colonialism is beyond the scope of this book. In 1901, the six British colonies united to form the Commonwealth of Australia. One of the first pieces of legislation passed by the federation's Parliament was the Immigration Restriction Act. The stated purpose of the act was to place certain restrictions on immigration and to provide for the removal from the Commonwealth of prohibited immigrants. As a policy, it was a clear attempt to regulate boundaries: a border is foundational to establishing a nation. Heyman and Symons (2015, 544) argue, "The border is a symbol of sovereignty, territorial polity and its role in various imaginaries of outside versus inside."

As Australia grappled to differentiate itself from its colonial masters, the themes of exclusion and inclusion came to the fore. Jupp (2007, 6) notes that Australia is "the product of conscious social engineering to create a particular kind of society." At Federation, a majority of Australians were Anglo-Celtic. This demographic hegemony was viewed as desirable—ideal even. Culturally, Anglo-Celtics were imagined to be a naturally coherent community with shared values, norms, and moralities. Other races threatened the ideals that held the Federation together (Neumann and Tavan 2009). In order to maintain a racially exclusive country, the policy curtailed groups deemed undesirable, economically threatening, or morally corrupt (Fitzgerald 2007). The ideal migrant came from the British Isles, with Europeans and white South Africans also acceptable, while people from Asia and the Pacific Islands were not encouraged to migrate. Acting attorney general, Alfred Deakin (1901, 4805), addressed the Parliament in 1901:

> *The prohibition of all alien coloured immigration, and more, it means at the earliest time, by reasonable and just means, the deportation or reduction of the number of aliens now in our midst. The two things go hand in hand, and are the necessary complement of a single policy—the policy of securing a "white Australia."*

Deakin's statements and the related policies illustrate that "border regimes treat peoples differentially, a diversity that is shaped by and affects moral thinking" (Heyman and Symons 2015, 543). In 1904, while people from the British Isles (Britain and Ireland) were actively encouraged to migrate, 7,500 Pacific Islanders originally brought over to labour in the sugar industry were deported. These parallel approaches led to the Immigration Restriction Act's notorious title: the White Australia Policy. While the specific wording of the act is not overly racist (Jupp 2007), the execution of the dictation test provided the space to regulate Australia's boundaries in a highly selective fashion. The dictation test required that prospective migrants have a command of a European language (or languages) of the examiner's choice. The test could be administered multiple times and in multiple languages, a policy that raised the standard of admission rather than giving potential immigrants additional opportunities to pass. Immigration officials, in effect, had exceptional freedom to reject non-Europeans. Of the 805 people who sat the test between 1902 and 1903, only forty-six passed (Yarwood 1958, 25).

These efforts were effective at regulating the population of Australia and reflected the exclusionary prejudices of the time (Jupp 2007). In the first census conducted in 1911, less than 1 per cent of those born overseas came from areas outside the British Isles, and the bulk of those immigrants arrived from Canada, South Africa, and the United States (Australia Bureau of Statistics [ABS] 1911, 116–119). Ten years later, Australia was less diverse than at Federation. Gradually, however, due to international events and domestic concerns, these practices were abandoned in favour of less racially restrictive norms.

After World War II, Australia began to hold a greater international role independent of the United Kingdom. It was a founding member of the United Nations and one of the eight nations involved in drafting the Universal Declaration of Human Rights (Morsink 1999). While internationally the country was at the forefront of inclusive social ideals, exclusionary policies persisted, particularly in the arena of migration and minority rights. The gap between the international image and domestic reality needed to be mediated. In 1958, strong domestic and international criticism of the dictation test led to its retirement (Palfreeman 1967; Jupp 1995). In 1964, the United States outlawed race-based discrimination, and the United Kingdom passed a similar act in 1965. The momentum of the civil rights movement reached the global stage: in 1965, the United Nations developed the International Convention on the Elimination of All Forms of Racial Discrimination. Prime Minister Gough Whitlam called for an end to exclusionary migration policy: "We say unequivocally that there must be no discrimination on grounds of race or colour or nationality" (Whitlam 1985, 498). The Australian government ratified the Racial Discrimination Convention in 1975, which gave rise to the Commonwealth Racial Discrimination Act of 1975. This act heralded a new era in Australian immigration policy. No longer concerned primarily with race, immigration policy became a means of selecting desirable people based on their potential economic contribution: skilled migrants were favoured (Hugo 1992). A points system that rewarded education, particular skills, or potential to conduct business meant a sizable number of migrants coming to Australia from the 1980s onward were middle class (Collins 2013). They joined a bourgeoning global and professional middle class (Stubbs 1996) who have accumulated a skill set "that will facilitate their positioning, economic negotiation, and cultural acceptance from different geographical sites" (Ong 1999, 18). These emigrants, though still largely from the countries that traditionally supplied migrants, also included new groups from wealthy areas in Asia. These latest groups tended to have higher levels of education and trade qualifications than the broader Australian population, and this appeared to translate into higher incomes (Inglis and Wu 1992). The exceptions to this trend were refugees who experienced "a massive loss in occupation status," regardless of qualifications and levels of education (Colic-Peisker and Tilbury 2006, 203).

Humane Regulation

At its core and since Australia's inception, there has been a preference for the controlled movement of people into Australia. The humanitarian program was no different (Dauvergne 2005). In Australia, as the previous section illustrated, immigration was largely policy driven. In contrast, refugee policy was initially reactionary, with the response tailored as the situation arose. While refugees articulated some aspects of broader immigration policies, they also represented distinct ruptures. During World War II, Asians and Pacific Islanders fled to Australia. After the end of the war, most refugees returned home. However, some—particularly those who married Australians—wanted to stay in the country (Neumann 2015). These refugees were gradually removed from the country, and those refusing to leave prompted the drafting of the 1949 War-Time Refugees Removal Act. A federal election in 1950 led to a softening policy on wartime refugees from outside Europe. In 1950, of the 853 refugees who could be deported under the act, 832 were allowed to stay (Neumann 2015).

Parallel to the deportation of Asian and Pacific Island refugees, the Australian government was participating in an ambitious resettlement program of European refugees. Through the International Refugee Organization, the UNHCR's predecessor, Australia resettled 171,000 refugees between 1947 and 1954. After the UNHCR was founded, Australia became a signatory to the 1951 Convention Relating to the Status of Refugees. This convention protects people seeking asylum and provides a legal framework for refugees. Australia's status as a signatory "means that Australia has legal and moral obligations to fulfilling the terms of these instruments" (McMaster 2002, 280). While it might appear that resettling European refugees was predominately driven by humanitarian motivations, Neumann (2015) argues this is not the case. Rather, the refugees coming from Europe were selected based largely on the hope they would bring skills to Australia. In order to maintain its economy, Australia needed a growing population that British migration was no longer providing. The acceptance of refugees was tied to the development of the nation and helping refugees was secondary in importance. Economic need, rather than political or humanitarian ideologies, provided the justification for accepting large numbers of European refugees. What Neumann overlooks in his analysis is the desire for Australia to be involved in the international arena. Accepting refugees buttressed Australia's image as both economically prosperous and an international actor independent of the United Kingdom. Refugees represented an asset in international relations while also providing a boom for the economy. Before the 1950s, the political nature of the refugee seemed relatively insignificant, yet the international credibility from resettling refugees was paramount. This focus began to change in the 1950s as various consular officials from the Soviet Union defected to Australia. The 1956 Olympic Games in Melbourne saw the defection of a further fifty-six athletes from the Soviet Union—most from Hungary. These defectors were deemed political refugees and accepted into Australia with little popular or political protest. Australia accepted fourteen thousand Hungarian refugees between November 1956 and December 1957 (Jupp 2007). These defectors transformed the image of refugees, morphing them into political symbols of the Cold War: "the struggle between the forces of Evil and Good" (Manne 1987, 233). The acceptance of these defectors crystallized a new image of refugees as not only economically capable but politically acceptable.

In 1977, increased arrivals of Indo-Chinese asylum seekers prompted the government to introduce a clear refugee policy rather than respond to events on an event-by-event basis (Mence et al. 2015). Australia was entering into a period of economic stagnation: the fleeing refugees were viewed as a potential economic drain. Global political powers, such as the United States, pressured Australia to accept the refugees, maintaining they were an extension of Cold War politics. Thus, accepting refugees from Vietnam was marketed by the government to the larger population as supporting political refugees while also providing humanitarian support. Again, the imagining of refugees was evolving. The refugees fleeing war in Vietnam were still considered political refugees, but humanitarianism began to emerge as motivation for acceptance and support.

In response to the arrival of asylum seekers, in 1977 the Department of Immigration created the Humanitarian Visa Program. This program was designed to protect those already legally in Australia through the onshore (asylum seeking) component and to resettle people offshore (refugees) deemed in need of humanitarian assistance (Phillips et al. 2010). It was a program similar to broader immigration policies in Australia in that it functioned to regulate the movement of people. In turn, this underscored the validity of national borders.

Australia is not unique in attempts to regulate its borders. Globally, migration from the Third to the First World, both legal and illegal, leads to fretful discussions about national security (McAdam 2013, 435). Schuster (2003) and Squire (2009) have observed an increase in suspicion of asylum seekers in the United Kingdom. Fassin (2005) argues convincingly that, in France, access to asylum was restrained, yet the country continued to provide humanitarian assistance to refugees in camps. In the United States, asylum seekers were subject to detention, particularly if they were fleeing countries that were considered to present a security risk (Welch and Schuster 2005). While subjecting some asylum seekers to detention, the United States remained the largest government donor to the UNHCR (UNHCR 2015). These tensions and contradictions appeared more generally as well. In 2000, the United Nations General Assembly passed two protocols regarding the movement of people: the United Nations Convention against Transnational Organized Crime and the accompanying Protocol against the Smuggling of Migrants by Land, Sea and Air. These protocols were developed with humanitarian aspirations, coupled with practical concerns. In effect, these protocols strengthen the legitimacy of strong border protection measures, potentially limiting people's right to seek asylum. Troublingly, "particularly on the part of the major destination countries, attempts to counter trafficking and smuggling seem to be motivated by a growing intolerance of all forms of irregular migration" (Gallagher 2002, 25). This focus suggests the border has a powerful symbolic and moral quality. Those who upset this moral quality are subject to suspicion.

Asylum seekers are presented politically and popularly as coming "to Australia for their own personal benefit rather than for humanitarian reasons"; namely, they are seeking economic opportunities (McKay et al. 2012, 129). Asylum seekers are considered to be exploiting Australia and posing a threat to Australia's culture, and are not perceived as genuine refugees (McKay et al. 2012). Assumptions of hidden motivations have led to asylum seekers being described as illegal migrants or as queue-jumpers. This labelling process results in asylum seekers becoming constructed as inherently different, a deviant social group that threatens national security and national identity (Pickering 2001). These concerns surfaced during the 2013 Australian federal election. The conversation was particularly charged regarding the role of each respective political party in protecting national borders. Tony Abbott, representing the Liberal Party—which would ultimately win the election in coalition with the National Party—campaigned strongly on his party's ability to protect the country's borders and "stop the boats" (Abbott 2013). "The boats" refers to asylum seekers attempting to reach Australia by sea to access their right to protection through the onshore component within the Humanitarian Visa Program. The Abbott Government actively fostered the image of a dichotomy between good refugees and bad asylum seekers. While this fieldwork was being conducted in 2012 and 2013, the division between these two categories became particularly pronounced. From a policy perspective, the overseas component became prioritized: "Beginning in 2013–14, the programme is being refocused to ensure that priority for places is given to people overseas entering as part of a planned process" (Department of Immigration and Border Protection 2013, 4). Genuine refugees emerged as synonymous with protracted humanitarian situations—living in refugee camps and coming through a planned process (i.e., the UNHCR referral system). These refugees became acceptable due to their experience as contained or managed subjects.

During my research, as noted earlier, there was a clear attempt to differentiate between the refugees and asylum seekers by the major political parties. A 2015 poll—roughly a year after these policy divisions became promoted—found 49 per cent of Australians agreed that "asylum seekers should be allowed to stay in Australia if they are found to be genuine refugees" (Essential Media Communications 2015, n.p.). However, the vast majority did not believe those arriving on boats were genuine refugees. A United Nations survey conducted in Australia found that two-thirds of Australians felt sympathetic towards refugees, particularly ones arriving from refugee camps (Gordon 2012). In contrast to these positive attitudes, the Lowy Institution Poll found 71 per cent of Australians agreed that the government should strongly deter people attempting to seek asylum by turning boats back towards the country of origin (Oliver 2014, 3). That same poll found 42 per cent of people agreed that "no asylum seeker coming to Australia by boat should be allowed to settle in Australia" (Oliver 2014, 10).

Though these polls have limitations, they hint at the broader argument put forth here that accepting some refugees allowed Australia to reconcile a strict migration policy. While attempting to exclude asylum seekers, Australia also desired to be viewed as a humane, righteous country. Australia was willing to accept genuine refugees from camps. In this context, refugees, though outside their nation, theoretically did not challenge the validity of national borders because they were contained in camps. Further, the Department of Immigration and Border Protection prioritized those deemed most vulnerable by the UNHCR. This resulted in an interesting cross-section of the population categorizing the elderly, disabled, single mothers, and key political leaders who had been tortured in one category. This category was linked by the assumption that all these people needed additional assistance, either due to social structures or to the presumption they were traumatized.

In an era defined by national boundaries becoming ever more selectively porous, who constituted a "victim" became progressively essentialized (Ticktin 2011). Trauma emerged as a significant barometer on the moral hierarchy of victimhood (Fassin and Rechtman 2009). Trauma is significant because it (inaccurately) sets refugees apart from asylum seekers and thus legitimized their admittance into Australia. Experiencing trauma, being traumatized, and needing assistance to deal with traumatic events have become cornerstones in assessing humanitarian assistance. By defining refugees as traumatized, Australia's moral standing was enhanced. The nation accepted the beneficiary, regardless of this assumed hindrance. Trauma was often invisible, and thus any refugee could be considered as slightly less mentally stable, slightly more vulnerable—in short, slightly less than a complete citizen. This hinted at the perception that the traumatized refugee, regardless of appearances to the contrary, was never quite on equal footing with untraumatized Australians. This more humanitarian focus created quite different imaginings and expectations than earlier understandings of refugees. Far from the economic skills European refugees brought, or the political solidarity of Soviet (and to a lesser extent Vietnamese) refugees, the attribute of the new refugees was helplessness.

Refugees were no longer considered economically proficient or politically capable but were perceived as dependent on the charitable impulse of Australia. In turn, this affirmed the notion of the nation as "good, prosperous and generous" (Dauvergne 2005, 4). Assisting some refugees absolved the moral ambiguity that the policy of offshore detainment, mandatory detention, and the act of turning back the boats necessarily presented. By creating a hierarchy of deservedness, and then assisting the most deserving refugees, Australia's humanitarian actions functioned "as the mirror in which the nation seeks a reflection of its benevolence" (Dauvergne 2005, 5). Jupp (2007, 182) argues further that accepting refugees served to foster Australia's status in the global community through the performance of charity. The acceptance of some refugees arriving through very specific channels was framed as an ethical act. This not only effectively downplayed the legal obligations of the nation to protect asylum seekers; it helped buffer the Australian brand from potential criticism both domestically and internationally (FitzGerald 2019). The end of 2013 saw the Minister for Immigration and Border Protection Scott Morrison (2013a) inform asylum seekers of Australia's expectations regarding their behaviour once in Australia:

You must not harass, intimidate or bully any other person or group of people or engage in any anti-social or disruptive activities that are inconsiderate, disrespectful or threaten the peaceful enjoyment of other members of the community...If we are going to release people into the community who have arrived illegally by boat from very different backgrounds, language groups and cultures with no prior exposure or connection to Australian society, we should at the very least explain what is expected of them in terms of their own behaviour, and be prepared to remove them from the community if those expectations are not met.

This information reached beyond the specific audience of asylum seekers but was more than realpolitick. Morrison's discussion of expectations of what makes asylum seekers problematic was telling: different backgrounds, languages, and cultures were a potential threat to Australia. More troubling, Morrison suggested that disruptive behaviour—conceivably political protest or requests for additional support or recognition—becomes an act of deviance when performed by people of different cultural backgrounds. This example highlights the role of borders in relationship to Australian national identity, but also, crucially, unity. In Australia, unity is discussed in terms of egalitarianism— everyone is equal, living in a classless society, and enjoying similar opportunities. Yet, as evident when discussing Australia's history in terms of immigration policy, for most of Australia's history, "egalitarianism emphasised exclusion of others considered unfit to participate in the process of nation-building" (Greig et al. 2003, 188). Immigrants, women, and Indigenous Australians are all sources of national *insecurity* that precludes them from full participation in the nation. The rhetoric of egalitarianism, in these contexts, functions to erase the reality of inequality. In this context, the state is often engaged in managing those who experience degrees of exclusion through policies of "immigration selection, through the constitution of ethnicity as a part of politics and through multicultural policy" (Pettman 1995, 78–79).

The Poetics of Inclusion

Once people are admitted into Australia, particularly if they are from a non-Anglo background, they begin interacting with a set of policies designed to integrate them into the nation-state. A policy of multiculturalism was adopted by Australia in the 1970s and reached its zenith in the 1990s (Jupp 2007). It has been defined in Australian policy in the following way:

All members of our society must have equal opportunity to realise their full potential and must have equal access to programs and services.

> *Every person should be able to maintain his or her own culture without prejudice or disadvantage and should be encouraged to understand and embrace other cultures. [The] needs of migrants should, in general, be met by programs and services available to the whole community. But special programs and services are necessary at present to ensure equality of access and provision. Services and programs should be designed and operated in full consultation with clients, and self-help should be encouraged as much as possible with a view to helping migrants to become self-reliant quickly (Galbally 1978, 1–2).*

Multiculturalism is a series of government policies that strives to celebrate diversity and combat racism. A key aspect of how the government approaches diverse cultures in Australia is through the premise of ethnic communities and funding flows through these groups. This has historic precedence: post-World War II refugees were expected to form ethnic community groups to ease the transition into Australian culture. Ethnic communities had an important role in providing social support (Colic-Peisker 2009; Westoby 2008). Thus, although they were partially bureaucratic constructions, ethnic community associations were not necessarily just the reflection of a government-level conflation of ethnicity, religion, and nationality. These groups functioned as an additional social safety net and fostered a sense of community for recently arrived members. A sense of community could lessen the stresses of resettlement (Lewis 2010). Regardless, ethnic community was a fraught category; ethnicity was often directly linked to a nation-state left behind. This paralleled the state-centred settlement ideal and promoted the "myth of the national family" (Arendt 2013 [1958]). In reality, there was often very little community in terms of social interactions, or even a sense of primordial oneness. Yet, for administrative purposes, the group was an ethnic community. This created an expectation for newly arriving groups to behave in a particular fashion, displaying certain community values and well-being and headed by a natural "leader":

> *Public agencies, politicians, and the media search for "leaders" of "ethnic communities" in the apparent belief that these are tribally organised with recognisable and generally accepted chiefs. Nothing could be further from the truth (Jupp 1984, 187).*

What emerged were quasi-communities that learned to engage with the discourses of expectations headed by specific ethnic leaders. This approach minimized all the very real fissures, politics, and potentially longstanding animosities within groupings condensed into ethnic communities.

Hage (1998) argues further that events designed ostensibly to celebrate multiculturalism in effect strive to contain and manage diversity. New arrivals are encouraged by service providers to showcase their ethnic backgrounds, particularly through the mediums of dance, crafts, music, or food. These displays suggest a positive desire to provide a space to celebrate difference. However, as Hage (1998, 87) argues, it represents a "strategy aimed at reproducing and disguising relationships of power in society or being reproduced through that disguise. It is a form of symbolic violence in which a mode of domination is presented as a form of egalitarianism." These are benign and nonthreatening displays of difference presented largely for the consumption of a white audience (Duffy 2005; Hage 1998). This selective representation obscures the fact that minorities are still relegated to the political fringe, normalizing social hierarchies that set the terms of participation for less powerful groups. It also overlooks the sense of obligation to perform difference in a way that fosters "good feelings of the nation." (Povinelli 2002, 6).

Even considering these insightful critiques, Australia has made remarkable strides from being "the most British" country in the world towards "the most multicultural" (Jupp 2007, 1). Castles (1988, 1992, 1995) observes that, though Australia's post-war immigration program was designed to keep the country white and British, it ultimately resulted in one of the world's most diverse societies. Further, Hugo (1986) argues that, a mere ten years after the introduction of the multicultural policy, Australia was one of the most ethnically diverse countries in the world. Again, this transformation took place due to international pressure and domestic agitation. In terms of government support, multiculturalism has since been experiencing a slow decline (Inglis 2009), punctuated most recently by the demise of the Ministry of Multicultural Affairs in 2013 (Department of Social Services 2014). The ideals of multiculturalism "in Australian public debate and reflected in the policy and rhetoric of both major parties" are being replaced by a neo-assimilationist ideal (Nolan et al. 2011, 670). While multiculturalism may be falling out of favour, its effects have been profound: Australia is demographically and culturally vastly different from what it was in 1973 when the first official multicultural policy was introduced. However, the shortcomings identified by Hage (1998) became particularly acute for refugees. The assumption of trauma and social disintegration further relegated these new groups to the social fringe, allowed to perform but not to fully participate.

Once in Australia, refugees began encountering "differing expectations for refugees upon arrival…which are manifested in the services offered and the pressure that refugees face to become self-reliant" (Fanjoy et al. 2005, 20). They received assistance through several service providers, both governmental and nongovernmental, to facilitate their eventual self-sufficiency. This was achieved through (but not limited to) English language classes, support with obtaining housing, general counselling, and health care, in addition to assistance in gaining employment. Considerable emphasis was placed on becoming integrated into the mainstream vis-à-vis becoming economically independent (Department of Social Services 2012). However, Australia took a measured approach and recognized the very real challenges new arrivals faced by providing sustained financial assistance, particularly in comparison to the United States and Canada (Fanjoy et al. 2005). This was largely due to Australia's more robust social welfare program rather than a particular generosity towards refugees. It must be further noted that the financial support was virtually identical to the amount an Australian citizen in a similar situation would be entitled (Department of Social Services 2015). However, prioritizing vulnerable refugees resulted in a high number of refugees dependent on social welfare.

Successful resettlement remained a crucial goal of the government. This goal was multifaceted and not limited to economic participation. Technical and Further Education (TAFE) provided 550 hours of free English tuition for refugees, and activities provided by other service providers tended to centre on acquiring spoken English skills. Successful resettlement was also understood to include obtaining citizenship, sending children to school, and attempting to fit into the broader Australian population (Fanjoy et al. 2005). At a 2013 citizenship ceremony, Minister Scott Morrison (2013b) stated,

> We will celebrate our democratic values, equality, and respect for each other and what unites us as Australians. We encourage anyone who is eligible to formally become a part of our community as Australian citizens.

The themes of democracy and egalitarianism were further highlighted in the *Australia Values Statement* refugees received before arriving in Australia during cultural orientation conducted by the IOM (Australian Government Department of Home Affairs 2020). Once in Australia, they received the study guide, *Australian Citizenship: Our Common Bond*, to prepare them for the citizenship examination (Commonwealth of Australia 2020). Democracy constituted the belief system to which new citizens pledged their allegiance during their citizenship ceremony: "From this day forward, I pledge my loyalty to Australia and its people, whose democratic beliefs I share, whose rights and liberties I respect, and whose laws I will uphold and obey" (Department of Immigration and Border Protection 2015). Thus, refugees such as the Bhutanese found themselves in an ambiguous position in Australia: as refugees they were incapable and passive, but as future citizens they needed to be capable and active.

Arrival Statistics

Australia committed to accepting five thousand Bhutanese, and between 2008 and 2013 they represented the fourth-largest offshore/humanitarian resettlement group (Department of Immigration and Border Protection 2014). The largest population (approximately 1,500 of the five thousand) of Bhutanese resided in South Australia. South Australia, in 2011, had 289,316 migrants within the broader population of 1.64 million (ABS 2011). There were some smaller communities of Bhutanese in Queensland and Tasmania. South Australia, and Salisbury in particular, has become a destination for some newly arrived Bhutanese. Though refugees do not get to select their initial destination city, those resettled in other areas in Australia viewed South Australia as the ideal destination for higher-caste Bhutanese. This is partially because the group has gained support from the local service providers and the Salisbury Council. The Salisbury Council was not a refugee resettlement agency but provided multiple services (English speaking lessons, art programs, and social outings) to the Bhutanese in the council area. The Salisbury Council, due to the demographic concentration of Bhutanese in the council area, had a central role in the social life of the refugees.

The Bhutanese in South Australia were concentrated in the City of Salisbury (referred to hereafter as Salisbury), a suburb with a population of 7,551 approximately twenty kilometres north of the Adelaide city centre (ABS 2011). Salisbury, along with its sister suburb Elizabeth, had ambitious beginnings. Although Salisbury was organically founded as a farming community in 1843, it became an auxiliary suburb for purpose-built Elizabeth. Construction of Elizabeth began in 1955 to provide housing for the employees of the automotive and manufacturing factories. Modelled after British "new towns" springing up around London, it was expected to be largely self-contained, with shopping, industry, and social facilities independent from the Adelaide city centre. Elizabeth was strongly marketed to potential blue-collar British migrants, described in housing trust videos distributed overseas as "a place to grow." It was a striking aspiration:

a decent house and a garden for working class people, who quite often couldn't get that in other Australian cities. They tended to have the working class down here in the flat and the professionals up on the hills... the segregation wasn't so severe (McCarthy 2005, para. 17).

These marketing endeavours were successful to the extent the city held the nickname "pommie-land." Elizabeth and Salisbury, as "British industrial districts," had their own form of segregation: until the 1970s, houses were not available to non-British (Jupp 2007).

This period of utopia was short-lived. By the 1980s, Elizabeth had entered into a period of decline, mirroring broader economic shifts away from manufacturing. By 2014, the few remaining factories continued to downsize as the automobile industry dissolved. Elizabeth's unemployment rates topped 33 per cent, one of the highest inner-city jobless rates in Australia (Department of Employment 2016, 29). Salisbury, adjacent to Elizabeth, followed a similar life cycle. Unemployment plagued the area, averaging over 16 per cent compared to the Australian average of 6 per cent (Department of Employment 2016, 6 and 29). Average income was lower than the broader South Australian average, as were overall education levels (ABS 2011). Salisbury and Elizabeth were some of South Australia's most violent suburbs (ABS 2014). All of these statistics impacted on the perception of the North Suburbs as violent, undesirable, and socially deviant, as viewed by the broader population in Adelaide. The service providers working in these regions, and particularly the council workers, were tasked with grappling with suburban decline, social marginalization, and classism. Salisbury was a cash-strapped suburb, and in 2012 the mayor publicly offered to sell council lands in order to attract much-needed funds to rejuvenate an outmoded town centre. While grappling with a more general suburbanization of poverty (Goode and Maskovsky 2001), the town aspired towards gentrification and cosmopolitanism. This was transformation linked to the increasing ethnic diversity of the suburb and broader, national aspirations towards multiculturalism.

Upon arrival, the Bhutanese were assisted by a plethora of groups including the local councils, organizations focusing specifically on refugees or migrants, and associations that stemmed from multicultural projects. Others focused on helping women or facilitating projects geared towards the youth or elderly. The nongovernmental organizations included the Migrant Resource Centre (MRC), Survivors of Trauma and Torture Rehabilitation Services (STTARS), and the Australia Refugee Association (ARA). Others, somewhat more peripheral but highly valued by participants, included Multicultural Foodies, OzHarvest, and the Penguin Club. While these were technically nongovernmental organizations, they relied heavily on government grants. Federal and state governmental organizations such as Centrelink, South Australia Health, Housing South Australia, and TAFE featured prominently in their lives. This multiplicity of social mediators had diverse affiliations and expectations, although most received their funding through the government. Grants were scarce and competitive—providers who had overlapping service profiles found themselves in competition.

Some organizations had clear time frames regarding how long they would provide support. For example, ARA provided leadership and development programs for humanitarian entrants who had been in Australia for less than five years (ARA 2014). Nearly every service provider came from a non-English speaking background, though they all spoke fluent

English as part of their jobs. They represented various migration backgrounds—some arrived as children from Europe in the late 1950s, some from Vietnam in the 1970s and 1980s. Others came to Australia via arranged marriages. These experiences and personal understandings, in turn, coloured their interactions with the Bhutanese. Despite the diversity of the service providers, they seemed to share general expectations of the refugee community: specifically, the ideas that male refugees could pose a problem, the group would be quite helpless, and they would need considerable guidance on how to become "Australian." The Salisbury Council also held the strong belief the Bhutanese were exceptionally spiritual, and that this constituted both a social asset and a potential problem. This, as the following chapters illustrate, had several consequences. It was through this expectation of spiritualism that refugees became tasked with the role of exotic other. Spirituality and exotic otherness were two examples of the affirmative Orientalism that the group adopted as a central aspect of their performative identity in Australia (Fox 1989; King 1999). The attitudes of service providers and council representatives were significant: they provided the Bhutanese with a social window into the broader Australian population. It was through these people, fleeting interactions with representatives of the government, and day-to-day life in their new suburbs that the Bhutanese learned what was expected of them as refugees in Australia.

Conclusion

Though there is an almost omnipresent narrative of egalitarianism and a set of multicultural policies to help realize that ideal, Australia has multiple social hierarchies that define who belongs and who does not belong in the nation-state. The Bhutanese, once the pawns of Bhutan, Nepal, and India, are incorporated into Australia's broader political agendas. The Bhutanese have presented Australia and the suburbs they settle in with a chance for redemption from harsh immigration policies. As refugees, they represent the humanitarian heart of a broader migration policy that is premised on exclusion. Rather than highlighting the legal obligation Australia has to protect refugees, the moral aspects have become central. The powerlessness of refugees has developed into their crucial attribute. They have passed a moral barometer in order to gain entry into Australia, but once in the country, their cultural norms and values have again become regulated and scrutinized. In Salisbury, they offer to the suburb a promise of both cosmopolitanism and community solidarity with which a dying city desperately wants to associate. This, in turn, has influenced their interactions with broader members of Australia.

Humanitarian ideals consistently reinforce the centrality of the nation. There is a necessary bureaucratic managing of people within the politics of humanitarianism. Those being managed are willing to participate—to a point. The following chapters explore the way resettled Bhutanese in Adelaide confirmed, pandered, and skirted the multitude of expectations they encountered in Australia to further their unique aspirations. The Bhutanese refugees were well aware of these hierarchies and attempted to work within them. They were versed in the most current debates, discussions, and potential consequences (positive or negative) of these dialogues. Although they were eager to integrate, there was also a desire to maintain cultural integrity and identity (Ager and Strang 2008; Feldman 2007). Heavily institutionalized relationships combined with very personal processes to birth a particular collective "Bhutanese-ness."

6
Domestic Humanitarianism

DURING AN INTERVIEW with the director of the Migrant Resource Centre in Adelaide in 2012, I asked why she thought so many Bhutanese were settling in Salisbury. The director straightened in her chair and proceeded to clarify that she did *not* encourage the Bhutanese to settle in the northern suburbs. She expressed concern that their settlement in this area would lead to a "ghettoization" of the Bhutanese. This possibility was, understandably, far from a desirable resettlement outcome. Her statements provided my first impression of Salisbury and highlighted the widespread perception of the suburb as socially backward and even morally corrupting. As vulnerable refugees living in this "dysfunctional" suburb, the Bhutanese required management, guidance, and intervention to avoid becoming ghettoized.

Salisbury is a marginalized suburb of Adelaide, struggling to reinvent itself. Due to their status as vulnerable refugees, the Bhutanese have become an asset for the Salisbury Council: just as the Bhutanese are a means for Australia to present a virtuous image as a nation, for the local council the ability to manage the Bhutanese appropriately promises redemption from a reputation of social backwardness. As vulnerable refugees, the Bhutanese have a worthiness that is a scarce commodity in Salisbury. The successful resettlement of the Bhutanese has been an opportunity for the local government to "prove" its capacity to host refugees. In turn, it hopes this enhances its social standing within the broader Adelaide, providing an alternative discussion to the more common conversation regarding the decline of the suburb. The resettlement of the Bhutanese is a model of domestic humanitarianism. The Bhutanese have the potential to embody "the nation's aversion to its past misdeeds, and to its recovered good intention" (Povinelli 2002, 18). However, to achieve this the Bhutanese needed additional guidance. As an ethnic community, there were clearly defined parameters regarding acceptable cultural practices and unacceptable cultural problems: both required government stewardship. Service providers and government officials described cultural problems to me as pre-existing norms or values that had the potential to fetter successful settlement in Australia. These included behaviours that were perceived as being harmful to women or at odds with Australian values in general. The relationship between men and women was the juncture that frequently came under scrutiny as needing reforming. It created a clear role for the Salisbury Council as the stewards of the Bhutanese. There is a degree of power in this position, and it resulted in competition over who "manages" those deemed deserving. The ability to manage refugees successfully reflects strategic uses of power embedded in the humanitarian system of values. While these transformations were discussed as necessary to emancipate women and promote egalitarian values, these changes had broader implications. For the Bhutanese, resettlement promised the long-awaited possibility of self-management. However, they found resettlement frequently became an extension of the humanitarian paradigm.

Binding Communities
Australians living outside the Salisbury Council area (some whom I met through their roles as service providers, others during exchanges in public spaces such as libraries or local transportation, as well as engaging with Adelaide-based academics) were quick to dismiss it as a racist, intolerant suburb full of people unwilling to redeem themselves. These perceptions characterized the problems the suburb faces (high unemployment, poverty, and crumbling infrastructure) as unique to Salisbury rather than due to broader political, social, or economic issues. There was little commentary that identified the suburb's decline as a consequence of the collapse first of the local agricultural industry, then of the automotive industry in neighbouring Elizabeth. As well

as the contracting employment market, there was a steady decline in state funding for bus shelters, street lighting, housing assistance, the public library, and bike paths, to name a few (City of Salisbury 2014). Yet these were endeavours that fostered livability and public safety. One of the few areas to remain robustly supported at the state and federal level was the Crime Prevention Program. Further, in 2015, Salisbury received two separate grants for the installation of closed-circuit television cameras; combined, this amounted to the largest grant awarded in South Australia (Attorney-General's Department 2015). These funding decisions both reflected and contributed to the dominant perception of Salisbury—security measures were necessary to regulate the unredeemable.

Rather than "a shared feeling of empathy or a moral imperative of solidarity" (Fassin 2012, 30), the suburb was largely stigmatized. As Australia reshaped itself economically, those who found themselves marginalized were imagined as inevitably ending up in that position (Peel 2003). In regard to Salisbury, it was concluded that the occupants of the suburb brought about its "inevitable" descent into poverty. Ethically, it became a space that was unworthy of broader egalitarian ideals. This helped obfuscate the political responsibility of those beyond the local council area, further relegating the suburb to the social fringe. Though faced with a shrinking body of financial support and a dismal reputation, the Salisbury Council consistently tried to mirror mainstream values of egalitarianism and multiculturalism. It had not given up on the possibility that its council area could redeem itself. The means of the council's transformation, the following sections argue, was linked to its ability to govern new arrivals and, most crucially, refugees.

There was a myriad of ethnic groups in Salisbury. I met people from Vietnam, Guatemala, Burundi, the former Yugoslavia, and Sudan, as well as persons from India, Nepal, England, New Zealand, and Italy. In May 2013, these diverse groups converged at the local council offices to participate in the Harmony Day celebrations. The multicultural event included singing, dancing, cultural costumes, and the presentation of a carefully made quilt—to be hung in the council's office. A different community created each of the quilt panels to symbolize an aspect of their culture. Some groups painted a map of their country, others a symbol of the nation they left. The panels were joined together yet at the same time remained contained within their specific region of the quilt. The Harmony Day events concluded in the singing of an abridged version of the well-known song, "I am Australian" (Woodley and Newton 1987).

The key ideal of public performances hosted by the Salisbury Council was one of bound, coherent ethnic communities as the vehicle for organizing difference. In Australia, as in the camps, "community" again emerged as a central strategy to organize and govern new arrivals. Communities became imagined as a "bureaucratic apparatus of political administration and control" (Rose 1999, 169). Thus, these formal promotions of multiculturalism were an important affirmation of governmental capacity. The Salisbury Council gained credibility by appropriately managing these diverse groups. However, Hage (2002) and Peel (2003) have both noted that the role of the government—local and otherwise—in managing multiculturalism may be overstated. Both authors have argued that suburbs experiencing an influx of new arrivals may be more tolerant than government and media narratives allow, and that government policies of multiculturalism may not be the glue holding these suburbs together (Hage 1998; Peel 2003). Nevertheless, the way the local council managed different groups did impart clear lessons for the Bhutanese. The expectations of community coherence were perhaps the most obvious. They learned that, as in the camps, community is institutionalized as a sector of the government. A clearly delineated community became a crucial mechanism to claim recognition (Rose 1999). There were also subtle lessons regarding selective inclusion and exclusion.

Directing Difference

The Salisbury Council hosted a Bhutanese-speaking seniors group[1] on Thursdays and Saturdays in a hall leased from a local church. Though technically the group's focus was on the elderly, every age group

---

[1]. The Home and Community Care Program (a Commonwealth and state government project) and the City of Salisbury jointly fund the program. See more at http://www.salisbury.sa.gov.au/Our_City/Community/Seniors/Cultural_programs#sthash.CZ6kWMYO.dpuf.

was represented—attendance on Thursday frequently reached two hundred. A member of the Salisbury Council facilitated the group. Representatives from service groups, government institutions, and social welfare groups frequently attended. Council staff generally invited guests, though there was one guest speaker (a Hindu religious leader from India) who was invited by members of the Bhutanese group. These guests spoke on a variety of topics such as healthy eating, the importance of exercise, how to access different support networks, and upcoming events in the area. Rarely did a week go by without guests in attendance. These people were introduced to the Bhutanese beforehand as "coming to see what we do," giving the impression that funding or the program's longevity depended on being able to show what the group does. Thus, the guests performed a dual function: to impart knowledge and witness the Thursday meetings. In response, the meetings took on a decidedly theatrical tone.

A semi-communal yoga class consistently opened the Thursday meeting (though if guests were not in attendance, this took place in a room adjacent to, rather than in the front of, the main hall). The smell of incense drifted above the chanting of the yoga participants. Women bustled about preparing vats of sweet tea seasoned with black pepper to be served after the morning activities. Fresh flowers—roses, marigolds, or daisies—from participants' gardens were placed in small vases on the makeshift desk the facilitator sat behind. Those not interested in yoga began their craft activities at a trestle table adjacent to the facilitator. Men congregated towards the back of the room, playing cards. It was a hive of activity that stimulated all the senses. The visitors filtered through the Thursday meeting, taking in the spectacle while the facilitator glowed with pride. The image the Bhutanese presented was very exotic and heavily influenced by the council representative's expectations.

The meeting was explained as a space for the Bhutanese—not Nepalese or Indian nationals with whom they may have ties—to practise English. Again, this underscored the importance of a cohesive community defined by national boundaries. For the Bhutanese, historically, these boundaries have been malleable. In the camps, this malleability became increasingly problematic for resettlement. Marriages across national boundaries were a severe hindrance to resettlement. In

Australia, the national boundaries of Bhutan became re-emphasized. As the previous chapters argued, refugees affirm the sovereign ability to regulate their borders (Schmitt 1985). Organizing refugees along national boundaries further normalizes this arrangement—again giving nation-states precedence over other forms of social organization. In doing so, ethnic communities based on national affiliations become naturally legitimate, effectively obfuscating alternative orderings.

The need to perform for visiting guests was equally a reflection of the council's capacity to govern, as it was the participants' ability to participate. The council representative went to great lengths not only to showcase the Bhutanese cultural exoticism but also the council's central role in supporting that culture. This became acutely evident during the creation of cultural crafts. Various art initiatives undertaken by this facilitator were designed to highlight the group's culture. The group initially used the supplied canvases and paintbrushes to paint flowers or practise the English alphabet. While these could be celebrated as evidence of steps the refugees were taking to master the English language, the facilitator was frustrated that the product was not sufficiently cultural. It was the hope of the facilitator that these paintings could be used in the council offices to highlight the ways the local government was supporting multiculturalism. The participants had to be consistently coached to paint things from their past, things representing their culture. Eventually, they produced landscapes of mountains and images of an agrarian lifestyle, and the facilitator evaluated these as appropriate demonstrations of their culture. The Bhutanese had a deep sense of gratitude to the facilitator for the services she coordinated for them and demonstrated sincere affection for the kind-hearted woman. I never heard a negative word about her, though participants did express a desire for more independence and control over community funding. However, given the precarious nature of project funding the Bhutanese experienced in the camps and in Australia, participants also were aware that it was important to please their benefactors.

The Bhutanese, as newly arrived others from underdeveloped nations, were understood as primarily rural and uneducated (Buchowski 2006). Yet, in terms of the Bhutanese broadly, the Bhutanese in Australia were disproportionately well educated and skilled. This was partially due to the resettlement policy that prioritized refugees who were victims of torture. Australia hosted several of the well-educated key leaders of the protests that led to exile. Additionally, many of the participants served as priests in Bhutan. Others were government agents or teachers rather than farmers. Most of this group, in contrast to the vast majority of Bhutanese refugees, had little experience of plowing a field. Yet it was the educated leaders who produced these agrarian paintings in the Salisbury Council's art initiatives. The artwork reflected the population deemed deserving in the contemporary humanitarian paradigm. Rather than political activists or skilled refugees, the Bhutanese were moulded into a popular imagining of refugees: marginally skilled and practising subsistence-based livelihoods. These imaginings could be interpreted as a guiding measure with very real repercussions: "regulating humanity entails creating humanity—the humanity of oneself and the humanity of others" (Barnett 2013, 385). Here, at the behest of the council staff, refugees were creating an image that supported an idealized understanding of a "Bhutanese refugee."

Even with these contradictions, activities were undertaken with singular gusto. The Bhutanese participants were eager to meet the facilitator's expectations. The facilitator and the participants both hoped this culturally unique art would be exhibited at the local council. This potential exhibition was understood as a means of elevating the status of the program and participants. Though the Thursday meeting was ostensibly for Bhutanese, it was also for the council. The Salisbury Council was actively contesting its own reputation. The facilitator, and the council more broadly, had a vested interest in illustrating their effectiveness at appropriately managing the Bhutanese. Across the many groups in Salisbury, the Bhutanese emerged as a uniquely publicized and valorized group. They were consistently invited to "showcase" their culture at events in the Salisbury area. This elevation of status related to their willingness to publicly perform what the council staff viewed as "good" refugee behaviours—an exotic community orientated towards virtuous values. Faced with few economic hopes for transformation, the council pinned its hopes for gaining credibility by appropriately managing the Bhutanese. It promised "a new moral contract, a new partnership between an enabling state and responsible citizens, based upon strengthening the natural bonds of community" (Rose 1999, 186).

The local council had a clear role in producing humanitarian subjects. Honing the lessons learned in the camps, the Bhutanese (re)presented themselves as a community deserving recognition by reflecting certain hallmarks to maintain what was a tenuous, and perhaps fickle, acknowledgement.

Just as Australia gained international credibility by accepting refugees, the council gained domestic credence by welcoming them. The people working for the council, many of them relatively recent immigrants themselves, consistently demonstrated a deep affection for and sincere effort to support the Bhutanese. Similarly, the Bhutanese refugees demonstrated a kind of kinship-bond towards the key service providers. Both parties entered the relationship with good intentions, yet they moved within a larger social framework that was, in many ways, inherently constraining. The Bhutanese offered a chance to show greater Adelaide that Salisbury was a multicultural suburb with capable (and benevolent) managers. This helped position them ever so slightly closer to the centre, shifting into a more powerful role as benefactors. Rather than being a problematic council area requiring either external management for anti-social behaviour or charity due to its relative poverty, Salisbury emerged as an area that provided support for those in the most need—refugees. This domestic example highlighted broader power dynamics within the humanitarian framework: compassionate gestures can also be politically advantageous.

> *If compassion emits this kind of public signal and has become something of a status symbol, then it might also unleash a competitive dynamic. Once a norm of compassion becomes solidly rooted, then actors will compete to demonstrate who is the most compassionate and attempt to avoid giving the impression that they are not (Barnett 2013, 385).*

Globally, nations vie for the status of being appropriately compassionate (Dauvergne 1999a). Just as there are implicit humanitarian hierarchies that determine who deserves to be alleviated from suffering globally (Fiddian-Qasmiyeh 2014; Fox 2001), these hierarchies also find traction domestically. The finite power of compassion creates a dynamic in which groups must compete against each other. For the suburb of Salisbury, the compassion quota had been largely exhausted. As deserving refugees, the Bhutanese had access to a language of compassion that others did not.

For the Salisbury Council, resettlement of the Bhutanese became a credible cause worth championing, "which implies not only neglecting other ones but also constructing them by choosing the best way of representing the populations assisted" (Fassin 2012, 226). This underscored the persistence of humanitarian hierarchies but also their selectivity. The selection of causes or groups deemed worthy not only precluded some from accessing opportunities for recognition but also effectively regulated those who were granted the privileged role. The council struggled into the mainstream by highlighting its ability to effectively create and manage spaces of cultural difference. It was responding to a constraining local environment where economic transformation was unlikely, but caring for and shepherding deserving refugees—manifesting Australia's humanitarian heart—was possible. The Salisbury Council actively used humanitarian hierarchies in an attempt to transform its reputation, a powerful demonstration of the way humanitarian ideals are incorporated into domestic governance. In turn, this mirrored broader processes that narrowed the scope of who deserved recognition and compassion: the hierarchy of deserving became acute on the margins of society. In Salisbury, this was further obfuscated by the very complicated problems of economic exclusion and social marginalization of the poor. While the Bhutanese represented a group deserving of compassion, there were some within the group who were considered more deserving than others.

Caring for Women
In Australia, the council staff member that worked closely with the Bhutanese told me she had to retrain the group regarding gender roles (Australian female, age 40s, interview, Salisbury, 2012 and 2013). She recollected that, when the Bhutanese group first started meeting, the women would sit in the back of the room while the men claimed the seats in the front. This was, in her mind, unacceptable and at odds with egalitarian aspirations. She worked to change the way men and women physically positioned themselves in the room. While women still walked behind the men when going to a social gathering, in the council-mediated space, women became the primary focus. During

resettlement, women were the targets of numerous activities. Through the local council and service providers, the government facilitated public speaking courses, educational opportunities, and social clubs catering specifically to women. Women were frequently invited to represent the Bhutanese through dance, song, or cooking at various social events. They arrived early to events, changing into saris from their more pedestrian daily wear. Groups of women fiddled over safety pins, the placement of *tikkas*, and the number of bangles for each wrist. Makeup was carefully applied in anticipation of singing and dancing, and even standing in tidy rows sent excited energy through the group. For cooking events, the women would meet days before to prepare the requisite items. Sharing stories about the camps was common as women used new technology to prepare food or justified why their traditional methods are still used. Women served these "cultural foods" at public events graciously, beautifully dressed, and mindful of presenting themselves as ambassadors for their resettled community.

Spaces for participation in the broader Australian population reflected a feminization of the image of refugee observed more generally: women (and children) had become representative of the ideal refugee due to their perceived helplessness (Malkki 1996; Hyndman 2010; Hyndman and Giles 2011). Women were encouraged by service providers to speak at events, perform their culture, and be the public face of the Bhutanese. The Bhutanese women were well aware of their representative role and took pains to ensure the public image they projected mirrored the expectation of service providers. They dressed the part and performed accordingly. The most photographed Bhutanese were two women in their eighties, both illiterate, who looked simultaneously exotic and vulnerable. They rarely attempted to speak to the visiting guests who wanted to share a photograph, but the two elderly women never failed to smile graciously. I cannot recall any visitors requesting a photo with individual men.

Though women were encouraged to be the public face of the Bhutanese, this did not reduce the difficulty of resettlement that the women faced. Their English skills were low in comparison to the men, and the social networks developed in the camps partially disintegrated due to resettlement. However, the feminization of the figure of refugee, though problematic, had opened up spaces for the women. In turn, this opening of spaces, however marginal, emphasized the attributes of an idealized refugee, "an ethical ideal of a politically blameless self, untainted by compromising political allegiances or economic self-interest" (Pupavac 2008, 276). The emphasis by service providers on making women the public face of the Bhutanese "acts as lessons on which aspects of refugees' identities are to be recognized and which have to be suppressed as a precondition for acceptance" (Szczepanikova 2013, 30). Women grasped these lessons, viewing the showcasing of their culture and "refugee-ness" as a means of repaying the support they received in Australia. They not only conformed to, but also attempted to exemplify, the role of ideal refugee. One of the reasons Bhutanese women may have been more willing to perform the roles expected of them relates to the traditional family structure.

Large, extended families with anywhere from three to four generations living under one roof (or in one instance spread between three neighbouring houses) are the norm among Bhutanese in Adelaide. At a practical level, this made finding suitable accommodation in Australia difficult as most houses are designed to cater to a nuclear family. In an ideal Bhutanese family arrangement, the parents provide for their children until they marry. Once they marry, the senior male child will remain in the family home. His wife joins the household, contributing to the running of the household while the son contributes economically. For parents, their efforts are partially reciprocated when their son brings a wife into the household. Ideally, wages flow up the family structure to the senior men and women. These can be parents or grandparents and occasionally older uncles or aunts. These senior members are the stewards of the family; decisions move down the family structure. They determine the family's livelihood strategies and social arrangements (such as marriage), and the redistribution of money. This is considered the model family. There are complications in its expression, particularly when marriages occur outside caste groups, when personalities clash, or as children attempt to assert undue social authority. Regardless of complicating manifestations, this idealized arrangement provides a reference point for the Bhutanese. The task of the parents is to care for or guide children, perhaps long into adulthood. A forty-year-old Chhetri man who is married with children explained, "You know, it is this way. Still I answer to my father" (interview, Salisbury, 2013). Another male informant explained to me, "I am thirty-three and under some guardianship" (Brahmin male, interview, Adelaide, 2014). The parents maintained most family affairs. Women were under the "guardianship" of their husbands, their husbands' parents, and other senior household members.

When discussing the social support the government provided, young men who were still studying, and women of all ages, often framed it in terms of "love" towards their community. One young man enrolled at university explained, "Australians must really love us because they care for us!" (unknown caste, age 20s, interview, Adelaide, 2013). The role of the Australian government in supporting the Bhutanese was subtly reframed to fit into existing expectations of family. For some, the government had become an integrated member. The government morphed into the role of "parent," caring for the children. It was socially bound to care for the Bhutanese until they could, in turn, reciprocate. It was crucial that this care be returned in the future. Accepting this support from the government was explained as being contingent on the younger generation "repaying" Australia through their future involvement with the broader Australian community. Generally, this would be through skilled employment or political engagement. Nearly an entire community that aspired to be recognized as competent, successful, and independent found itself very much dependent on the patronage of the government. Few Bhutanese were satisfied with this arrangement and actively tried to renegotiate that relationship away from dependency. The first response brought the Australian government into an existing family framework of obligation. Women and young adults generally relayed this understanding. The second shift, actively promoted by men, created a role in Australia based on their role as community managers. Confronted with the prospect of ongoing dependency, some Bhutanese attempted to transform the role of the government-as-benefactor, while others transformed their refugee-as-beneficiary role. These understandings highlighted the dearth of opportunities to contribute that the refugees, particularly men, faced.

Council staff and service providers described men to me as wanting to dominate the public space and attempting to silence women. Female refugees, on the other hand, became a group that needed to be given a voice. They were allies to the local government, while refugee men became problems who threatened gender relations. The creation and maintenance of this divide, while providing additional spaces for women, also led to cleavages. One council staff member consistently wanted to "bring domestic violence out in the open" and encouraged women to publicly confess, during community events, when their husbands were abusive (female Australian service provider, age 40s, interview, Adelaide, 2013). When this did occur, this council employee saw this as an affirmation regarding her ability to create a liberating social space for women. This is not to suggest that creating a public space to discuss and transform social issues is inherently negative. However, as in the camps, domestic violence was presented to the group as a template with pre-delineated contours. This had the effect of excluding violence that did not conform to expectations, such as women-against-women violence, which did not fit easily into the stereotypical "male-as-problem" model. Attempts to create women as a coherent group, with gender transcending all other forms of difference, were problematic. Further, these interpretations facilitated and naturalized control over women. Liberating female refugees ultimately reconstructed them as a category in need of exceptional guidance. This overlooked not only their existing capacities but also very profound variations. The women, like the men, were far from a homogenous category with the same experiences or aspirations.

The Critique of Dowry and Polygamy
The Salisbury Council's attempts to emancipate women occurred in conjunction with an evaluation of the group's pre-existing norms and values. These norms were often intimate, personal, and political. At an everyday level, the Bhutanese became subject to critique and evaluation that led to "unacceptable" norms becoming problems council staff or service providers needed to fix quietly. Norms such as polygamy and the exchange of dowry were flagged in the camps as inappropriate and unacceptable. After resettlement, the persistence of these behaviours required attention, intervention, and possible eradication by the service providers and the local council. In essence, difference that was deemed morally unacceptable needed regulation. The UNHCR in Nepal functioned as a supra-government, governing at a distance through moral regulation and reconfiguration. A robust education in

international values strove to mould the refugees into a very specific humanitarian subject. The language of protection, community, and compassion obscured this political process of resocialization: "humanitarian governance justifies acts of interference on the grounds that they are necessary to save lives and reduce human suffering" (Barnett 2013, 389). In Australia, governance became intimate as local government representatives again constructed the Bhutanese in an image that furthered the former's political goals. This produced what Fassin (2007) describes as "aporia": apparent inconsistencies that illuminated the tensions embedded in humanitarian forms of governance. Government requests for refugees to change and transform were presented as a means to help women—fulfilling egalitarian aspirations. The reluctance to recognize this as a form of governance functioned to depoliticize the requirements for social reforms. This became acute when discussing norms outside the acceptable. The practice of polygamy and the exchange of dowry were norms flagged by several service providers as "cultural problems." Yet rather than dispensing with norms that could be considered a sign of unacceptable otherness, Bhutanese refugees reconceptualized them to fit into a narrative that was better suited for a broader Australian audience.

Polygamy was becoming less common in the camps due to several situational factors. People no longer had large farms that required considerable family labour to maintain. Men also faced fewer options for a well-paying livelihood that could provide the financial means of supporting multiple wives. Resettlement further accelerated the move away from polygamous marriages, but it also forced existing partnerships to be legally dissolved. Polygamous couples were not recognized as a valid family unit. Before resettlement, polygamous couples separated to expedite a process that could take several years. In Australia, there were few instances of polygamy. Though there may have been more families than I was aware of, it is unlikely they occurred much more frequently than the research could establish. While the Commonwealth recognizes a polygamous union that occurred overseas, polygamy is technically illegal in Australia (Family Law Act 1975, section 6). Once resettled, polygamous families may initially live in different cities, then move to reunite. In the large polygamous households in Australia, families rented adjoining apartments or houses in the same neighbourhood to be together. Regardless of these efforts to maintain their family structures, participants recognized polygamy was considered an inappropriate family arrangement in Australia.

The group was very concerned that public discussion regarding the practice of polygamy would undermine their status as deserving refugees. For participants, polygamy again became a site of tension: though encouraged to practise their culture, some very intimate social institutions were deemed unacceptable. Family is a central value in Australian culture (Dauvergne 1999b), yet it is a very specifically defined institution: largely nuclear. This is the family structure that is imagined to be naturally in line with egalitarian ideals (Kingston 1975). This highlights that alternative family structures may be understood as a threat to accepted, foundational social structures. Abolishing polygamy was framed as necessary for the broader public good: a means of protecting Australians from immoral norms.

Dowry was another "cultural problem" needing to be abolished in order to help the Bhutanese women. There were campaigns in the camps, led by NGOs, which depicted dowry as a social ill. However, it remained a significant social norm. In resettlement countries, the exchange of dowry is similarly rejected as a cultural crime against women or a social evil (Oldenburg 2002). The exchange of dowry was deemed illustrative of a broader social system that favoured males. However, it can have multiple roles. It can function as a social safety net, by ensuring women have access to some resources—often gold jewellery that can be liquidated if necessary. Dowry, for many participants, was evidence of the high status of a woman and one of the few pathways women had to acquire property. Failure to provide a dowry suggested the bride was not valued in her natal home. Perhaps even more significantly, a lack of dowry negated the groom's obligation to provide gifts to the bride's family. While the gifts to the bride's family were not as substantial (they included cooking implements rather than gold jewellery), the exchange showed dowry has a complex social role, helping solidify social bonds.

Participants were keenly aware that these alternative understandings of dowry were not considered legitimate and could undermine the moral legitimacy of the group. During conversations, women sought to distance themselves from the practice of dowry. Female participants speculated that a few conservative or "maybe some backwards people" still maintained this norm "from the past, the old way of thinking." They, on the other hand, were modern and righteous. The women I worked with adamantly denied that a dowry was paid to their husbands. Still, they all conceded that their husbands received a considerable amount of gold during the wedding negotiation process, and they brought ample amounts of jewellery into their new family home. This was not done by a compulsion, a point the women felt was significant, but rather was simply a very generous gift. This shift in the stated underlying motivation (gift versus payment) enabled the women—and the community because they were tasked with the role of representing the group— to maintain an image of progressive ideals and behaviours, minimizing norms and values that cast them in a negative light.

Abolishing polygamy and dowry was framed as a necessary measure to oppose patriarchy and the oppression of women. In short, they were seen as norms that directly undermined the ideals of egalitarianism. Framed this way, these transformations became a necessary means to protect women from a patriarchal value system. This was a clear moral evaluation regarding how a spouse was selected. The Bhutanese were acutely aware that cultural or ethnic difference was acceptable, even celebrated, but only to the degree that it did not deviate too far from mainstream norms. The group did not feel they were in a social position, even if they held Australian citizenship, to discuss in public these social institutions. Open discussion threatened to rebrand the Bhutanese: from good refugees to social deviants. Marriage arrangements were both moral and political, but approaching polygamy and dowry as problems that need fixing depoliticized them (Ferguson 1990; Fisher 1997). These became problems that needed regulation and fixing rather than being complex social systems requiring considered discussion. While framed as necessary transformations to equalize men and women, it was not a given these changes would reduce the suffering of women. For the women left behind in Nepal, when polygamous families had to decide whether to stay together in the camps or pursue a new life, these moral frameworks probably directly contributed to an increase in suffering. This highlighted the unintended consequences that occur when "an ethics of care meets the will to control" (Barnett 2013, 389).

Fitting into Australian culture was of fundamental importance for this group, and they were eager to maintain that their Bhutanese identity was highly compatible. Yet it was a complex process:

> *Political identities have to be worked out, negotiated, creatively compromised between peoples who have to or want to live together under the same political roof...and this coexistence is always grounded in some mixture of necessity and choice (Taylor 2011, 140).*

Here, norms excluded from the cultural space of inclusion became pushed further into the social margins. This dynamic again illustrated that "though the borders can be opened to some...getting through them still involves fitting into a very Australian mould" (Dauvergne 1999b, 44). Deeming some norms to be beyond the sphere of political discussion functioned to sustain the status quo. For the Salisbury Council, the ability to effectively convey the lessons regarding acceptable and unacceptable was a reflection of its ability to govern. Governing difference also entailed perpetuating values that defined the mainstream.

Though the council actively shaped, managed, and directed the Bhutanese regarding public events and—as this section illustrates—private behaviours, the staff did not see themselves as participating in a form of governance. This may seem paradoxical: the council *is* the local government. Yet the council staff I worked with did not interpret these "necessary" changes as a form of governance. Rather, they self-described as facilitators—helping or guiding those in need. This steadfast disavowal of governing illuminates broader power dynamics within the humanitarian framework and the ease with which politics becomes obscured by the language of compassion. The Bhutanese became both the "object and the target for the political exercise of power whilst remaining, somehow, external to politics and a counterweight to it" (Rose 1999, 168).

Managing Men

The ideal figure of the refugee became equated with the image of a vulnerable female in need of assistance. Service providers encouraged women to move out of the domestic arena and into the public or economic sphere. Some women found employment, particularly in areas that were an extension of their domestic roles—providing childcare or care for aging parents, cleaning, or working in vegetable cultivation and harvesting. Employment outside the home tended to be part-time and low-wage.

Bhutanese men, on the other hand, found themselves less than ideal refugees. Further, they found few options to move beyond refugee status. There was a profound dearth of full-time, well-paid jobs in Adelaide's northern suburbs. As women found themselves expected to be socially active and perhaps even employed, men became entrenched in the domestic sphere. One man in his early thirties explained his changing role:

> *I used to be a teacher in the camps, but here I cannot find a job. Normally, my wife would take care of the children, but she found a job—our neighbour helped her. Now, I volunteer, but I am mostly the house minister now. I take my girls to school and keep everything running* (Brahmin, interview, Salisbury, 2013).

For most men, this was a striking change from the camps, where they dominated schools as teachers and the camp's internal management structure. In Nepal, opportunities to work outside the camps were fairly common. In Australia, Bhutanese men still commanded a powerful role within the household. Though women performed the singing and dancing at public events, men actively regulated the image of the Bhutanese as a community. Yet Bhutanese men perceived a lack of available "masculine" roles in Australia. One woman explained her perspective on the men's experience: "Here, they say everyone can be empowered. But we know, someone has to give up a little" (Brahmin female, age 30s, interview, Salisbury, 2013). Several men who were farmers reflected that, before arriving in Australia, they aspired to own farms akin to the ones they had in Bhutan. Owning a farm promised self-sufficiency, autonomy, and status. In Adelaide, they did not think owning a farm would be possible due to cost, the urban/suburban setting, and the strikingly different climate. Others, particularly those with college degrees, hoped for employment proportional to their qualifications. While a few had been able to move into paid employment, these were viewed as exceptional accomplishments.

In Australia, refugee men staying at home with the children was not celebrated as a step towards a more egalitarian society. It was not recognized as the important contribution to society that it is. Rather, work in the domestic sphere, regardless if it was performed by men or women, did not appear to be considered "real" work by the government and nongovernmental actors the Bhutanese interacted with. This perspective may illustrate Ticktin's (2011) finding in Paris: the contributions refugees make to the economy are consistently minimized. Further, the part-time, low-paying jobs women found did not infringe upon an already precarious labour market. The Bhutanese were aware that even though there was a government policy in place to prioritize skilled migrants that could contribute economically to the country, Salisbury was not in the position to benefit from these prioritized migrants. Now, paid employment is not the only pathway towards social status, either in Australia more broadly or for the Bhutanese specifically. However, the men I worked with both in the camps and in Australia consistently spoke of the value of gainful, paid work. They used the colloquialism "eating another man's sweat," either through camp rations or social welfare payments in Australia, to articulate this was not considered a desirable way to live. This sentiment is similar to what Banki and Phillips (2017, 14) found in the Bhutanese refugee camp setting: dignity can be restored and fostered "by being given the opportunity to be contributing...members of society."

In the camps, men imagined that after resettlement they would have jobs capable of supporting their families. One male refugee explained his chance of contributing to Australia was lost due to his status as a refugee—the status was viewed as undermining his ability to fulfill his obligations to his family. This man was not particularly old, had the equivalent of a high school education, spoke English, and held leadership roles in Bhutan and in the refugee camps. He volunteered for a local resettlement organization and hoped to one day find gainful employment but did not think this was a realistic aspiration. He pinned his hopes on his daughter, who would outgrow her status as a refugee and be able to aspire towards being a contributing member of Australian society. He, on the other hand, found himself without a role beyond "a refugee the government is helping."

This particular experience elucidated a widespread concern among the generation of men roughly between ages twenty and sixty. For men, recognizing the government as an additional "parental" figure was an unsatisfactory way of making sense of ongoing dependence. They worried they would never be in a position to reciprocate, not due to a lack of capabilities but due to the prevailing expectations linked to their status as vulnerable refugees. While women were expected to expand their social roles through the assistance of service providers, men found their role to be increasingly confined to the "traumatized refugee."

Hutchinson and Dorsett (2012, 57) observed the funnelling of refugees in Australia into the category of traumatized: "New arrivals were routinely referred for specialist trauma counselling services. These routine referrals seemed to be based on an assumption of trauma." At a very practical level, in Australia, being a traumatized refugee is a recognized disabled status, converted into additional monetary support. One trauma counsellor explained his concern regarding this precedence:

> *The problem is, people come and they see how hard it is to find a job and then people [service providers and other Bhutanese] tell them there is extra support if they are traumatized. So they call and ask how they can get this benefit—that they are refugee and they are traumatized. They say, "Yes, I was tortured, I am from Bhutan, I have been tortured—where is my payment?" The problem is then they are stuck and it is even harder for them. For some men it is very shameful to get money from the government—if they had a way to earn it would be better. Here, they focus on the pain because they want people [Australians] to know they are here because they have suffered (male Australian service provider, interview, Adelaide, 2013).*

Noble intentions seem to drive the need to naturalize the relationship between refugees and trauma. However, during my research, this assumed relationship had the consequence of further alienation from mainstream Australia. While trauma can have a powerful legitimizing effect, it also reinforces refugees' status as primarily victims. Marlowe (2010, 195) found that the emphasis on helping Sudanese refugees by focusing on their traumatic experiences effectively regulated their ability to make "a meaningful contribution to society." This hints at the perception that the traumatized refugee may never quite be on equal footing with untraumatized Australians.

For Bhutanese men, trauma morphed into a dominant feature of refugee identity in Australia. In the camps, trauma was a peripheral concern. Though there were counselling services available, not a single participant I spoke with sought them. They were worried doing so would cause the locals to consider them incapable of employment, educational achievement, or acceptable social interactions. In Australia, trauma assumed other, potentially positive, aspects. Trauma provides recognition and speaks the language of compassion that the contemporary moral climate rewards (Fassin and Rechtman 2009). Refugees' trauma legitimized them and also translated into tangible monetary support or recognition from the government. However, trauma was also equated with a presumed lack of mental capabilities or social skills, and with being of questionable value to the nation (aside from illustrating the nation's charity). Many worried these associations had impacted on their reception and prospective degree of inclusion in Australia: "people won't recognize the skills that we are bringing, people just think refugees are poor people without any skills" (Chhetri male, age 20s, interview, Salisbury, 2013). Another expressed a similar concern that the refugee/trauma relationship minimized their employability: "the skills that we have will not be valued" (Brahmin male, age 40s, interview, Salisbury, 2013). Participants frequently voiced concern that while trauma helped people understand their journey to Australia, it also undermined future aspirations (Neikirk 2017). They worried they would always be viewed as negatively different to their Australian hosts.

The humanitarian focus on suffering, trauma, and help can have unfortunate consequences. Rather than integrating refugees into citizenry, it positions them as "poor people beseeching the state's benevolence...the obliged cannot assert a social right or demand precise rules: they must submit to the modalities imposed on them" (Fassin 2012, 78). Normalizing the image of the helpless female and traumatized male has the consequence of mediating their ability to engage with the broader population. By expecting widespread trauma, Australia effectively regulates a portion of the refugee population who are viewed as potentially problematic refugees—men. Because they are approached as impaired, they are not expected to participate in Australia. These constrained understandings result in defined power relations being framed as compassionate efforts to help (Neikirk 2017). This obscures the "power-structured relationships maintained by techniques of control" (Fisher 1997, 446). Trauma both regulates and relegates the male refugees. At a practical level, it normalizes their absence from an already crippled local labour market. Crucially, it curtails their potential for participation in other spheres—political or social. Trauma has emerged as a significant barometer of the moral hierarchy of victimhood, but it is also a measure of the receiving nation's moral credentials. The nation accepted refugees, regardless of this assumed debilitating trauma. In these circumstances, the allocation of social welfare may be functioning as a pressure valve, minimizing the risk of social protest or demands for political transformation (Piven and Cloward 1971; Halper 1973; Giddens 1985). Men were not passive in this situation, and many did attempt to wrestle a more publicly active role back from the resettlement agencies and local government.

Legitimate Managers

The concept of a Bhutanese refugee community was deliberately cultivated in the camps. Though the refugees were encouraged to be self-sufficient and self-regulatory, the camp system precluded this outcome. In Australia, refugees from the same nation, with few exceptions, were similarly categorized as a community (with tiers of "deservedness" embedded within). In such a situation, non-Bhutanese facilitators, council workers, and service providers made a considerable effort to create a coherent, strongly linked community of Bhutanese. These communities provided the cornerstone for organizing multiculturalism, promising access to resources and recognition. During resettlement, refugees adhering to this model gained the ability to self-regulate. Achieving this ideal was presented as obtainable in an undefined future, but before this point, refugees needed to be managed.

The Salisbury Council actively promoted itself as a legitimate manager of the Bhutanese resettlement. It provided the appropriate spaces for the Bhutanese to practise their culture and educated them regarding acceptable norms and values. In turn, this emphasized Salisbury's attempts to move beyond its status as a suburb that needed fixing to a suburb that had the capacity but required support. This was a clear and deliberate effort to claim positive recognition. It also articulated the hierarchical relationship embedded in the humanitarian framework between those who received and those who gave (Agier 2010; Barnett 2013). One reaped positive political rewards, while the other saw basic needs met. Yet this power dynamic was not absolute. The council's ability to manage refugees (and multicultural groups more broadly) was punctuated with a sense of insecurity. One council facilitator expressed uncertainty on several occasions regarding her ability to "do it" the right way. This self-consciousness was demonstrated in their reactions to my research.

The service providers based in downtown Adelaide or in one of the more affluent suburbs were satisfied with the fairly standard information sheet I provided to all respondents. Some expressed an interest in seeing the final ethnography. These service providers seemed confident that, even if my research were not glowing, their work would be little affected. Their capacity to perform the necessary work never seemed to be in question. The Salisbury Council, on the other hand, made frequent requests for specifically tailored letters, reviews of my research, and updates regarding what participants were saying. Initially, I thought this related to bureaucratic considerations. Then I assumed it related to strategies of governance: information could be used to manage the population more effectively (Hage 1998). While these may have been factors, the requests were best understood as a reflection on the council's status in relation to greater Adelaide. As a council area, it was subject to strong criticism because its populace was viewed as inherently dysfunctional. While the Adelaide-based service providers were quick to explain the council was doing "good work," the sense from the council staff was they did not have the luxury of a margin for error. Already pushed to the far fringe of society, they could not afford to mismanage new arrivals. The Bhutanese provided a means for the council to illustrate its righteousness, which had the potential to transform into improved political and social recognition. Perhaps unsurprisingly, the more powerful Bhutanese shared the same goal—they too wanted to shed stereotypes of incapacity and social backwardness in hopes of self-managing "their community."

The Bhutanese in Salisbury, headed by the Bhutanese Australia Association of South Australia (BAASA), responded to this ideal of community. BAASA is a democratic organization that provides support to new arrivals in Australia. Though lower-caste refugees started the association, they lost control of it as higher castes became the demographic majority in South Australia. During fieldwork, BAASA was the largest and most dominant Bhutanese-run group, not only in Adelaide but also in broader Australia. Its stated goals were to provide services for the Bhutanese community while promoting the ideals of inclusiveness and human rights. These goals mirrored the ideals of the UNHCR, whose stated purpose is to safeguard the rights and well-being of refugees. It also mirrored the Universal Declaration of Human Rights. The adoption of these models was perhaps a product of the refugees' long interaction with international donors in the camps and illustrated the manner in which globally circulated ideals permeate everyday life. It also illustrated the group's astute reading of which values carried moral weight and its strong desire to be viewed as global peers. While BAASA strove to be the representative body for all Bhutanese in Adelaide, the complex caste dynamics challenged its ability to regulate and manage the image of a unified Bhutanese community.

The members of BAASA, almost all of whom also volunteered for the City of Salisbury, started an ethnic school that ran on Saturdays. Participants frequently described the banning of the Nepali language in the classrooms of Bhutan as a "human rights abuse" that contributed to their eventual exile. This was further compromised in the camps, as the schools were English-based and most parents viewed learning English as a valuable opportunity for their children. As a result, there was a generation of children who could speak Nepali but could not write it. In Australia, the importance of everyone becoming fluent in English resulted in several people voicing concern that the culture they fled Bhutan to preserve would be gone within one generation. Consequently, the idea to start a school emerged. In addition to helping preserve the Nepali language, the school gave people who experienced difficulty securing employment (particularly older male participants) a means to maintain social status. From 2012 to the beginning of 2014, approximately eighty students between the ages of five and fifteen attended the school each Saturday morning.

In early 2014, the school became a contentious issue. Classes were held in a space provided by the Salisbury Council. The group started the school without adequately informing the council of its plans. The founders further attempted to skirt the council by securing funding from a separate government agency that specialized in funding ethnic schools. Once the council became aware of the school, after the school had been running for several months in the building the council provided for adult English classes, it also faced an insurance restriction that only covered the space for the use of adults. An inability to secure insurance for children to use the space led to the suggestion that the school might have to close. Eventually, the council was able to arrange a compromise with its insurance provider. Though a satisfactory solution was found, the matter was not immediately settled. The attempts by the Bhutanese to navigate the channels of bureaucracy independent of the Salisbury Council facilitator were met with a rebuff. The council worker explained the following to the school's founders in an exasperated tone:

> You need to go through me. You don't directly approach people, like the mayor or the council and tell them you are doing these things. They come to me and I don't know what they are talking about. There is a process here in Australia. I don't know what it was like in Nepal or Bhutan, but here there is a process. I have told you this before. You cannot just do these things (Australian female, age early 40s, 2013).

One of the key volunteers (and a driving force behind the formation of the ethnic school) responded, with darting, downcast eyes:

> Oh, oh yes. Oh, we are refugees. You know, our English isn't that good. We don't know—we don't understand these things. The system here, you know, it is different. Yes, oh, I am sorry. We don't know these things (Brahmin male, age late 30s, Adelaide, 2014).

The actions of the Bhutanese went against one of the basic tenets of expected refugee behaviour, as many of the participants understood it: refugees are not capable. To reclaim the secure position of refugee, they deliberately downplayed their obvious capabilities for the service provider. The refugee slipped into the role of passively grateful beneficiary. The volunteer quoted has impeccable English and is a skilled (and employed) translator from Nepalese to English. It is particularly interesting he slipped into a syntax that supported his claim of a lack of English skills. I do not believe this was an attempt at sarcasm. Rather, it was an effort to express deference to the council worker. He also appeared to recognize that an assumed lack of English capabilities was significant: it underscored his helplessness in Australia. Bhutanese leaders glided back into the patron/patronized relationship that the council facilitator needed them to maintain. Here, the slippage between compassionate helpfulness and political aspirations was manifest.

This incident suggests the Bhutanese were not completely satisfied with a contained relationship. In the camps, projects were largely dictated to them or selectively funded to promote a larger agenda. The actual day-to-day running of endeavours was largely independent of "outside" oversight. While international staff visited occasionally to provide further training, map outcomes, or perform an evaluation, they were not present on a weekly basis. In Salisbury, the council had a more intimate, everyday presence in the lives of the Bhutanese. It actively mediated the social spaces, ensuring the Bhutanese reflected well on them and learned the "right" behaviours. The above exchange, perhaps more than anything else, speaks to the council's fear of losing its role as managers.

The Bhutanese desired more autonomy but could not risk the loss of patronage. Though they were attempting to strike out on their own, they were still bound to their providers. As such, it was not their desire to sever relationships with service providers. Rather, they wanted to maintain those relationships while building their own parallel (and more autonomous) organizations. These efforts, though perhaps a perfect example of a community acting as agents on their own behalf and exercising the skills of self-reliance service providers claimed to be promoting, were not met with praise. Refugees were told they should be striving towards self-reliance, but concrete attempts to do so were deemed problematic: threatening the acceptable image of a deserving refugee. This further demonstrated that refugees played complex roles, not only internationally but also domestically. Being in a position to help can be powerful—as the dynamic in Salisbury illustrated.

As a microcosm of the nation, Salisbury strived to be welcoming, humanitarian, and the "right" kind of multicultural. While this would not fix the problems of unemployment or poverty, it challenged the widespread perspective that the suburb was "full of racists" and morally unredeemable. It provided the slim possibility that the problems facing the suburb might no longer be considered individual problems but rather part of a larger problem that could be addressed through policy. Salisbury's marginal status thus made it somewhat dependent on the Bhutanese. Though there were clear attempts to manage the Bhutanese, the council could not afford to question all their values. This could risk the Bhutanese, their "model community," no longer being willing performers. However, the central Adelaide-based service providers were in a different position. They had taken a firm stance that one of the group's central values—the caste system—required reforming. Service providers located downtown, due to more secure social standing, had more autonomy than the Salisbury Council. While the Bhutanese in Salisbury were able to negotiate with the local council representative to leave this issue (largely) alone, they were not in the position to negotiate with the Migrant Resource Centre. This led to a very different pattern of interaction.

Shifting Dynamics

BAASA and the MRC originally enjoyed a close relationship. Initially, the MRC director went out of her way to support the group by attending its events, writing letters of support on its behalf, and actively recruiting Bhutanese to work as volunteers. After a few years, however, there was a falling out. The director of the MRC explained why she decided to cut ties with BAASA:

> BAASA *maintains a very tight grip on power that bars women and lower-caste individuals from accessing positions of prestige within the community. They effectively silence certain members of the community while putting on a cheery "we are democratic, look how democratic we are" facade. I spoke at the big party last year to celebrate three years in Australia and then noticed that the lower caste was not allowed to eat—at a party to celebrate democracy! (Australian female, interview, Adelaide, 2012).*

The MRC did not forbid members of BAASA or the higher castes from attending events. Rather, it distanced itself from the political organization but continued to work with the *community* of Bhutanese. This is a vital distinction. Although community is a crucial political category in Australia, it is idealized as apolitical. Communities are imagined as beyond "any calculated assessment of self-interest" (Rose 1999, 177). Evidence to the contrary suggests the need for further education and increased efforts by service providers to recreate the "natural" community. After cutting ties with BAASA, the MRC took additional steps to support and facilitate Bhutanese whom they perceived BAASA was excluding—the lower castes. BAASA, in response, attempted to call the services provided by MRC into question by having Bhutanese request additional items of support. One request that elicited particular frustration was for electric rice cookers.

The basic goods provided on arrival in Australia included food, linens, white goods, and some kitchen equipment (Department of Immigration and Border Protection 2015). In the United States, where resettlement is more ad hoc and variable (Ong 2003), several Bhutanese families received a rice cooker either from a church group or a similar agency helping with the resettlement process. After living without electricity for decades, news of luxury items being literally given away to the Bhutanese in the United States spread quickly through the social grapevine. This led to the requests in Adelaide. The MRC interpreted this as a lack of gratitude for the support Bhutanese refugees received during the resettlement process, describing the group as being like a squeaky wheel—always demanding more. Requests for additional things, particularly material goods, were not considered good refugee behaviour.

The rice cooker had additional significance beyond being a simple luxury good. For the Bhutanese, rice is synonymous with food. A meal is not complete, nor even considered a meal, without it. Women, almost exclusively, prepare the food for the family. Much discussion surrounds a woman's ability to cook rice properly. It should not be at all crunchy but also should not be gluggy. It should hold together when mixed with cooked *dhal* (lentils) and *khargati* (vegetables) but not stick to the fingers when eaten. Ill-prepared rice elicits negative comments. Food, due to a long-standing lack of refrigeration and persistent cultural norms, must be freshly prepared for every meal. Rice is cooked a minimum of twice a day; attempts to recycle rice from lunch in order to save time are frowned upon. Rice must be picked through, washed, and cooked in the pressure cooker—with the cook staying vigilant for the necessary two whistles. A rice cooker promises perfectly prepared rice that can sit—warm and still fresh—until the evening meal. It can save the women of the household a tremendous amount of time.

Due to the central importance of rice, the MRC's failure to supply these items led BAASA to suggest it was not sufficiently aware of the group's needs. Most, if not all, participants could afford to purchase a rice cooker. However, the rice cooker issue provided a crucial means for BAASA to wrest power away from a service provider that was publicly critical of it. BAASA drew upon its power within the Bhutanese community to critique the services provided by groups with whom it did not get along.

To add insult to injury, BAASA lost its government grant shortly after the MRC (South Australia's largest migration organization) withdrew its support. While this may have been a coincidence, BAASA remained suspicious of the MRC. A BAASA member commented on this:

We have the spirit to give. Here smaller voluntary work does much good but it is not rewarded, now we don't have the space to give. Now, the government just wants to give to bigger organizations but they have so much bureaucracy. The smaller ones can work much more effectively. We want to help each other (Brahmin male, age 30s, interview, Salisbury, 2012).

By suggesting the Migrant Resource Centre had not provided a basic need for the group, the balance of power between benefactor and beneficiary subtly shifted. BAASA used this request to deflect attention away from its internal power dynamics by suggesting the MRC did not understand the group at a very basic and fundamental level. This, in turn, also undermined the MRC's critique of the caste system. These tense power struggles revealed that willingness to perform the role of an ideal refugee did have limits. Ong (2003) found Cambodian refugees in America similarly adept at shifting the power balance. While refugees become subjects within a value system with particular norms, "they also modify practices and agenda while nimbly deflecting control and interjecting critique" (Ong 2003, xvii). The Bhutanese were asserting themselves but also attempting to maintain, and work within, the framework of expectations.

During fieldwork, I frequently purchased supplies for the female-dominated Thursday craft session held for the Bhutanese in Salisbury. The women reciprocated these purchases with bracelets or necklaces. Every lady went to great lengths to clarify that the gift was a new item or handmade. Many left the tag on and took pains to draw my attention to it. Initially, I was reluctant to accept these gifts, assuring them they were not necessary or expected. Eventually, I stopped protesting and realized the transformation they were trying to bring about. They were attempting to establish our relationship as an exchange of equals. These women were not simply recipients but able to reciprocate.

The desire to reciprocate in a dignified manner (by giving a new item or a product of their labour) became crucial as the Bhutanese found themselves embedded in Australia's social welfare system. In the camps, though people were well aware of Australia's robust social welfare system, no one imagined they would find themselves dependent again. Rather, virtually everyone I spoke with anticipated the accoutrements of citizenship and the contributions they could make in their new countries. Upon arrival in Australia, they realized contributing was not necessarily expected, or perhaps even desirable. They found themselves still humanitarian subjects rather than potential citizens. In this regard, they are quite different than other groups of immigrants. An early model of refugee migration focused on the economic contributions new arrivals would make, but this expectation seems to now be reserved for those migrating through a skilled migration stream. In the contemporary climate, the role of the refugees who are resettled is to resolve the moral dilemma that global inequalities present.

Conclusion

Humanitarian ideals are expressed in a variety of ways: sympathy, a desire to help others, the need to guide the vulnerable. As agreeable as these ideals seem, they obscure unequal relations between those who require help and those who help. As politics becomes redefined by its incorporation of the language of humanitarianism, groups no longer credible within a humanitarian framework are further silenced (Fox 2001). In turn, this framework lends itself to the reproduction of power dynamics rather than their transformation. In Salisbury, the morality of those living in the northern suburbs was subject to scrutiny and evaluation. The broader South Australian community found them morally deficient. The Bhutanese, on the other hand, were a morally deserving group willing to work with the council. They conformed to the multicultural framework that naturalizes the role of the government as the mediator of cultural difference. Yet conformity in some arenas did not mean that a clear process of either domination or emancipation was occurring.

Attempts to manage the Bhutanese were met with mixed results: the council staff was not trying to create "robots programmed with cultural rules," but they were trying desperately to assert themselves as competent and capable managers (Abu-Lughod 1991, 158). The means of doing this was linked with the ability to reinforce Australia's national culture as morally apt and ensure the Bhutanese were not corrupted by the suburb nor were themselves a corrupting presence. This leads to uneven power relations. In turn, this articulates broader themes of social relations between those with the power to govern through compassion and those who are governed by compassion (Barnett 2012).

Refugees are made into an acceptable subject—both in refugee camps in Nepal and once resettled—through a series of interventions. Physical regulation and moral parameters have attempted to create a unique humanitarian subject: a Bhutanese refugee community bound by righteous ideals. In Australia, vulnerability and victimization are not only expected but the very hallmarks of an exceptional, and thus acceptable, refugee. Despite the benevolent sentiment, seeking the most vulnerable effectively regulates refugees in two ways. First, it makes the pool of deserving refugees smaller, whereby the status is "no longer a right but a prized status" (Zetter 2007, 188). Second, the suggestion persisted that vulnerable refugees are expected to rely on social welfare indefinitely. These understandings legitimized a strict border policy. In order to maintain Australia's egalitarian ideals, only a select, deserving few could be supported. Approaching refugees as "naturally" dependent becomes a means of both maintaining borders and regulating the few refugees who were granted sanctuary. This both "creates and guarantees the situation that the law needs for its own validity" (Agamben 1998, 17). This had the effect of creating an exclusive group of refugees who were kept aware of their "prized status." In turn, this minimized the possibility they would threaten the system that allowed them to enter Australia. While refugees remain a crucial political topic, they remain largely absent from participating in these political discussions. This absence is normalized by stripping them of their capacity to be political actors—depoliticizing them into helpless and vulnerable, traumatized and incapable humanitarian subjects. This compassionate understanding serves to contain and regulate the Bhutanese, effectively relegating them to the social fringe.

7
Sanitizing Otherness, Becoming Australian

ADELAIDE, bounded by the Mount Lofty Ranges to the east and the Gulf Saint Vincent to the west, sprawled north and south for kilometres. This low-lying urban space was anchored by a similarly low-slung downtown. Rising up from the city centre was Adelaide's tallest building—the thirty-one-storey Westpac House. Crowning the structure was a stylized "W" that resembled the *tikka* worn by higher-caste Vishnu Brahmins. Both men and women wore a *tikka* made of ash, sandalwood, clay, or coloured starch mixed with water on the forehead. The shape and colour of the mark provided visual information regarding one's position in the caste hierarchy and the associated level of purity. The lower castes in Adelaide drew parallels between the building's domination of the skyline and the higher caste's ability to control an image of a "Bhutanese community." While drawing this parallel, they were similarly quick to downplay its significance. One of the favourite jokes among the lower castes was to call Vishnu Brahmins "the Westpac-ers"—an attempt to ridicule the mark that functions to remind the lower castes of their polluting status.

The lower castes saw the Westpac building more frequently than the higher castes because they had moved to suburbs closer to the city centre, while the higher castes remained in the north. This is a reincarnation of settlement patterns in Bhutan, where the lower castes were expected to live away from the higher castes. In the camps, this was not possible, as huts were allocated without consideration for the caste system. There was little autonomy over where people lived and who their neighbours were. The higher castes found it difficult to exercise control over things that affirmed their status in the caste hierarchy. This included broad aspects, such as maintaining a monopoly on education and, at a more basic level, regulating who had access to water sources. In Australia, the perception that people had "freedom in our homes" allowed them to revive a system that was seriously compromised in the camps. A Brahmin man in his forties explained to me, "Some might feel unaccepted but most are obeying the law of Australia. You cannot stop people who say or feel this way, but in my mind we are cooperating with the law" (interview, Adelaide, 2012). Here, the man is highlighting an interesting point—while the caste system is considered deviant in Australia, it is not illegal. Further, there is no expectation that Caucasians with ancestral roots from the same country would automatically welcome each other into their homes. The expectation of community varied remarkably based on proximity to the dominant culture in Australia. For the higher castes, and particularly those of the Brahmin caste, reviving a caste system was part of shedding their status as refugees and returning to a ritually pure Bhutanese community based on a caste hierarchy. Many of the lower castes, on the other hand, were anxious to be free of the caste-based order and embrace the broader Australian notion of community linked to the ideals of egalitarianism. This was an understanding that liberated some and challenged others, providing pathways for particular kinds of action while blocking others.

The previous chapter argued that refugees play a central political role in local governance—not as actors but as objects of care. Underpinning these interactions was the assumption of a unified ethnic community—linked to a nation and further solidified by a common refugee experience. Part of the attraction of community "lies in their apparent naturalness: their non-political or pre-political status" (Rose 1999, 189). Further, community is buttressed through the assumption of "purity or naturalness, wholeness or wholesomeness of origins...communities, cultural traditions, or nationalities" (Malkki 1995b, 516). However, "cultural identities are in a constant state of flux and attempts to define them as statistical and administrative categories—for example, through the idea of 'ethnic communities'—are problematic" (Colic-Peisker and Farquharson 2011, 581). The resettlement process encouraged a particular model of

Bhutanese community that ignored social realities. It is through an examination of the caste system that "the difference between the imagined figure and its enactment" can be fruitfully analyzed (Ticktin 2011, 15). In this context, it is the imagined Bhutanese refugees who are an apolitical, naturally united community butting against the living and breathing Bhutanese who are politically organized and highly stratified. Competing notions of a Bhutanese community emerge: one based on the values of hierarchy, and the other on the values of egalitarianism. As the Bhutanese found themselves jockeying for support and recognition in Australia, caste again became a site of intersecting ideals regarding how the world should be ordered.

Caste as Community

Due to several decades of education about caste in the refugee camps and similar education programs in Australia, higher-caste participants recognized that caste is viewed as an inappropriate social stratification system. Yet, for the higher-caste Bhutanese, caste provided the central way of organizing virtually every aspect of their world: from what they ate and drank, to whom they married and the names of their children. After decades of compromise in the refugee camps, Australia promised the chance to become pure Bhutanese—a return to an earlier notion of community based on ritual hierarchy. For many Bhutanese of all castes, this understanding of community held great value. A community organized along caste lines was understood as a means of ordering their social world in which everyone has a clearly defined role. Abiding by these structured roles ensured the possibility of future positive reincarnations. As a closed system of social stratification, these potentially positive future reincarnations were the means of moving up the social hierarchy. In mainstream Australia, there was a similar emphasis on community but with a radically different meaning. Community, ideally, is based on egalitarian values, the capacity for self-determination, and the chance for everyone to have a "fair go." A "fair go" in Australia is a value related to enjoying equal opportunities, but it also links to the hope that the socially downtrodden can—and should—succeed. Of course, there are contradictions in the manifestation of these ideals.

Even with these shortcomings, these aspirations did have an influence on how an Australian community was imagined and what it should be striving towards. Caste, with its emphasis on stratification and restricted social mobility, was viewed as incompatible with the Australian ideal of community. Service providers targeted the caste system, due to the perception that it was at odds with egalitarian notions of community, as a "cultural problem." Service providers attempted to reorganize the group from a firmly gradated scale into an egalitarian community. Yet, for the higher castes, this was the first opportunity in decades to be wholly themselves—able to live away from the lower castes and reaffirm a social framework based on ritual purity. For the higher castes, balancing external demands with internal ideals led to distinct performances that publicly affirmed the rightness of broader Australian ideals while privately affirming social hierarchies. In June 2013, the Bhutanese responded to these competing notions of community by holding the caste-affirming sacred thread ceremony but rebranding it as an exotic interpretation of an Australian norm: the housewarming party. As a ritual, it affirmed the Bhutanese concept of community based on social hierarchy but was presented to an Australian audience as a nonthreatening cultural performance.

    The party was just beginning when I arrived. I sat with several neighbours and a few representatives from the Salisbury Council under a covered carport. The host informed me that he had invited many people from the broader Australian population, expressing disappointment that the mayor and other higher-ranking officials could not attend. The higher castes in Salisbury continually made efforts to engage with the mayor, hoping such a relationship would eventually lead to greater autonomy and social independence. Seats were procured for the visiting non-Bhutanese, while close to fifty Bhutanese participants stood. After a few minutes of chatting, the host's son appeared. It immediately became obvious to me that we were not witnessing a housewarming party but a sacred thread ritual—one of the most significant religious ceremonies in a male Brahmin's life. Before a Brahmin male reaches the age of ten, he is not bound by the rules of purity and pollution that dictate food consumption and personal interactions across castes. He can eat and play with anyone he chooses. When he dons the sacred thread during an elaborate ceremony, he is expected to live his life in line with the Vedic scriptures. He can no longer consume meat or eggs, alcohol and smoking are prohibited, and he must now avoid ritually polluting people. The three plain cotton threads are twisted to make one rope worn across the chest representing his unique link to God (Bennett 1983) and symbolizing his position at the apex of the caste system. It is only removed when he marries (to be replaced by a six-threaded rope to symbolize greater responsibility) or at the once-yearly purification ritual during which he receives a replacement thread. If a Brahmin chooses to disregard the Vedic teachings by consuming meat, drinking alcohol, or transgressing other taboos, he should remove the thread and is only entitled to wear it again after extensive purification rituals.

The ceremony around the sacred thread was aesthetically beautiful and highly theatrical. The young man was carrying a sceptre and was wrapped in a single piece of cotton cloth. His head was freshly shorn (aside from a small tuft of hair) and an exceptional *tikka* denoting his status as a Vishnu Brahmin stretched across his brow. He began collecting ritualized alms, and outsiders were given uncooked rice sprinkled with flowers to present as an offering. Singing, incense, cedar garlands, rice dotted with flowers, and the colourful attire of the smiling Bhutanese created an intoxicating spectacle. To the untrained eye, this was a unique and spectacular interpretation of a housewarming party. Yet all these acts stressed the structural component of the Hindu caste system. The boy was wrapped in a single piece of cloth because, according to the caste hierarchy, clothes are sewn by the lowest castes and hence polluting. Further, cotton is used for the cloth and the thread because it is considered purer than silk, which is an animal by-product. His shaved head had a small tuft of hair towards the back, representing not only his ritual cleanliness but also his unique connection to the gods. His *tikka* represented Lord Vishnu's (the supreme God's) footprint—again affirming the Brahmin relationship with the head of the Hindu Pantheon. The uncooked rice and flowers are given to bless the young man with knowledge, health, and purity. These two gifts are considered nonpolluting, and thus it is acceptable for lower castes (including Australians) to present them to higher castes. The sacred thread ceremony was about highlighting social boundaries. Here, the Brahmins were distinguishing themselves as not only different but ritually purer than other people. However, additional steps were necessary to balance dual aims: conducting a ritual that reinforced the stratified cosmic-moral order that elevated Brahmin males above all others while appearing to be an inclusive community.

A crucial ceremony in a young, high-caste Brahmin male's life was reframed for outside guests as an exotic performance of an Australian social norm. This performance was successful: the higher-caste Bhutanese presented a ritual as a cultural performance that did not test the fundamental expectation of Australian egalitarianism. Earlier events had not been so successful, and on those occasions council staff reprimanded the group for practising the caste system in a way that was viewed as discriminatory. Though alternative interpretations of this event are possible— such as not wanting to confuse Australians with an elaborate ritual—this explanation fits with broader patterns that occurred both in the camps and in Australia. Durkheim (2008 [1912]) understood rituals as fundamentally concerned with internal organization, and while this interpretation cut to the core of the sacred thread ritual, arranging an audience of outsiders suggested something more was occurring. Baumann (1992, 100) looked specifically at rituals in multicultural contexts. In these settings, rituals can be understood as "a claim to public attention, public space, and public recognition" and attempts to make a symbolic statement. The sacred thread was a potentially divisive subject due to its confirmation of the caste hierarchy. Neighbours and city officials might have been reluctant to attend a ceremony that affirmed the caste system, a hierarchy participants were told was at odds with the Australian ideology that everyone has the right to a "fair go." As such, the Bhutanese deliberately attempted to manage the image of the event to fit within Australia's version of community. Those involved in organizing this ritual staked an active claim to the creation of a Bhutanese identity in Australia. Involving council staff and bringing them into the high-caste network, even partially, was strategic. Presenting a ritual fundamentally about caste hierarchy as a cultural performance functioned to mask the persistence of the caste system. It was an exercise in displaying public boundary markers—clothes, music, dance, and food—in a "celebratory" multicultural framework (Hage 1996, 1998).

One young, high-caste man claimed that caste was not a major issue in the camps. He explained to me that "in the camps it might be thirty or forty degrees [eighty-six or 104 degrees Fahrenheit], but no one knows because in the camps no one talked about it, now it is an issue" (interview, Adelaide, 2012). This was a very different interpretation from what I saw in the camps, where there was widespread awareness of the caste system. Though there were many lower-caste people changing religion, and the higher castes could verbalize the "issues" with the caste system, there were few demands that the higher castes change. Part of this related to the form humanitarian governance took in the camps—it was largely from afar. In Australia, service providers were very involved with the Bhutanese and felt empowered to critique institutions. Rather than being an abstract issue the higher castes knew produced discomfort for some but could be minimized through knowing the "correct" answers to non-Bhutanese's questions, caste was pushed to the centre of conversations with Australian service providers. Many high castes tried to minimize that the caste system was, in fact, a real problem, and they did this by linking it so firmly to the organization of their community. The strength of appealing to the concept of community lies in its natural authority (Rose 1999). Naturalizing community as an organic institution, here the high-caste man is literally saying caste is as natural as the weather, makes it difficult to dispute—it exerts a moral legitimacy (Selznick 1994). However, in Australia a community defined by caste is viewed as unnatural: a social creation governed by clear rules and codes of conduct. The political construction of a refugee or ethnic community undermined its credibility (Rose 1999). The Bhutanese caste system called the inherent goodness of community into question not only because the behaviours appeared to contradict egalitarian ideals but also because the very basic imaginings of community were threatened (Selznick 1994). It was a community structure that articulated the inequalities and hierarchies that were largely hidden or publicly disavowed in acceptable notions of community. Thus, caste had to be performed in a way that minimized these contradictions.

While Australian guests were invited by the hosts to attend this ceremony, the lower castes were not. When I asked one Brahmin man in his early forties why he would eat food with me but not with lower castes, he explained, "We have been taught for thousands of years that the low castes are not good. It is hard to change these things just because someone says" (interview, Adelaide, 2013). Here, the almost contradictory inclusion by exclusion of the lower castes became clear. The lower castes were firmly entrenched in the purity spectrum, and in this way they were intrinsically tied to hierarchical notions of community. As an outsider, I was also integrated into the purity index but in a less absolute fashion.

Some higher castes would eat with me and accept cooked food or tea from my hand. Others would eat with me but not eat food I had prepared. A few would not eat or drink in my presence. Most participants approached Australians and Westerners more generally as being similar to an impure but touchable middle caste, although this was not a concrete position. I, like most Australians, was not fully integrated into the community hierarchy, and thus my level of purity was somewhat fluid. For the lower castes, their ritual pollution was well established, as the man stated, "for thousands of years." They were concretely integrated into the community hierarchy. This presented a particular intersection between competing notions of community. The Australians in attendance had a less definite role than the lower castes who were more integrated into the community hierarchy. As non-Bhutanese Australians were not full members of the community, our liminal status allowed us to be treated in a more egalitarian fashion—eating in proximity to the higher castes, taking part in their ceremony. Our lower degree of integration into the social hierarchy meant our presence was tolerated, while a fully integrated member of a similar status was not welcome at such an event. In turn, this illustrated that, although excluded from the event, the lower castes were actually considered fuller members of the community than the non-Bhutanese guests in attendance. Their role, in the eyes of the Brahmins, was not to witness ceremonies or share food. During this gathering, everyone could appear to be equal as we ate together and participated in a ritual. Within the caste-based social structure, it was expected the lower castes would be excluded, and on this basis a ritual that centred on social stratification could be presented to outsiders as affirming egalitarian ideals.

Though involving outsiders was strategic in terms of controlling the image of the group, it was also problematic. Despite the success of this particular performance, service providers were aware of the dual concepts of community. In Australia, service providers had an active role in transforming the Bhutanese from a community segregated along caste lines to a community bound by the ideals of unity. In order to bring about this transformation, the service providers had to be informed of internal dynamics. When they were not informed of the nuances of the cultural performances they witnessed, they frequently expressed a sense of betrayal. Internal dynamics that suggested the group was not united as an ethnic community threatened the carefully crafted patron–benefactor relationship the Bhutanese had cultivated in Australia.

The need to curate the image of the Bhutanese to outsiders was a recurrent theme throughout my research. The curators were almost exclusively high-caste males who were literate, bilingual, and eager to work with an outside researcher. When I first started working with the group, they suggested I use a written, rather than a verbal, consent method. I soon discovered that championing a written consent form was a strategy to contour and effectively regulate my research through seemingly technical means. Brahmin males, more than any other group, held the skills necessary to sign a consent form. For many other Bhutanese, written consent proved to be problematic; illiterate participants felt ashamed of their inability to sign their names, and asking them to sign with an "X" or other mark was humiliating for everyone involved. That the group had a sharp vision for my work demonstrated their well-honed understanding of the system of humanitarian values they moved within. They understood the need to convey particular sentiments, take on certain roles, and reflect specific values. My task, in their minds, was to further their cause, eliciting sympathy that could be converted into political might, and present them in a positive light. They also recognized that major rifts within the group regarding issues such as caste threatened to undermine the image of Bhutanese as a deserving community. The Brahmin men did not want it publicly discussed because its presence countered the dominant value system. Its persistence suggested an attempt to skirt service providers' reforming efforts. This partly reflected the expectations that the Bhutanese should willingly reform themselves to project the appropriate set of values. The following section analyzes the attempts by service providers to reform the caste hierarchy and the actions of some of the lower castes who took an active role in attempting to transform the group to mirror Australia's expectation of community.

Transforming Caste, Reforming Community

In Australia, as in the camps, there were different ethnic groups, religious affiliations, and languages subsumed under the label, "Bhutanese." In the camps, the higher castes (Brahmins and Chhetris) represented a demographic minority, yet they exerted considerable power and viewed themselves as the public face of the Bhutanese. In Adelaide, they had become a demographic majority and essentially monopolized the image of the Bhutanese in relation to service providers, the general public, and researchers. The public performances were strongly influenced by service providers' expectations. The Thursday social group (described in the previous chapter) welcomed newly arrived Bhutanese with a scarf-giving ceremony. Scarves, a symbol of goodwill, were commonly exchanged in the refugee camps when a person departed, began a new stage of life, or returned from a journey. In the camps, white silk scarves printed with the *ashtamangala* (eight auspicious symbols) were given on special occasions, such as weddings, departures, or arrivals in groups that were predominantly Buddhist. Yet I also observed Hindus practising this ritual in the camps, with woven cotton, rather than silk, scarves. As mentioned earlier, for many Brahmins, silk was considered impure. I observed Brahmins giving cotton scarves to those within their social strata but not to those below them. (The exception to this was the scarf given to me by my Brahmin research assistant's family, but this exception highlights my more liminal place in the caste hierarchy.) The physical contact necessary for such an exchange to occur between a higher caste and a lower caste would entail unacceptable exposure to ritual pollution. However, in Australia the desire to maintain a physical distance from the lower castes was deemed unacceptable by service providers and at odds with Australia's egalitarian values.

While the scarf-giving ceremony in Australia bore some similarities to those in the camps, these subtle distinctions appeared to have collapsed. The Salisbury Council's non-Bhutanese facilitator ran the events and oversaw the inviting of guests. She was adamant that the group not become "an elite Hindu club," and went to considerable effort to ensure that all newly arrived Bhutanese received an invitation to participate in the community event (Australian female service provider, age early 40s, interview, Salisbury, 2013). During one ceremony, new arrivals (including myself after returning from Nepal) stood in front of the group of roughly two hundred Bhutanese and accepted scarves as a symbolic gesture of goodwill. The Salisbury Council staff member did not present the scarves. Rather, the duty to present scarves to the new arrivals fell to the higher-caste volunteers who essentially monopolized the council's volunteer positions. This ceremony was intended to affirm an assumption that this was a community reuniting, that their nationally based identity bridged ethnic, religious, and caste differences. Some of the new arrivals were invariably from lower castes—some were Hindu, others had converted to Buddhism or Christianity. While the council member saw this as a chance to affirm Australia's notions of community, the acts expected to be performed by the higher castes ran counter to many of their notions of community. The ceremony became an elaborate, if somewhat awkward, performance. I had been working with the community for over year at this point and the unease between the higher-caste presenter and the lower-caste recipient was, to me, palpable. Such an exchange ran counter to sanctioned caste interactions. In the face of discomfort, the ceremony was performed, scarves accepted, and photographs taken by the council employee. The lower-caste arrivals were publicly welcomed to Australia, and an invitation to participate in the group was extended.

New arrivals, if they were from lower castes, and most acutely the ritually polluting castes, might attend the Thursday meeting in Salisbury once but then quickly realize the welcome ceremony was not a genuine invitation by higher-caste Bhutanese to join the group. For the higher-caste Bhutanese in Salisbury, balancing the demands of the council with their own value system required considerable social manoeuvring. They had to go through the motions of publicly inviting the lower castes to participate while privately visiting them and making it known they were not welcome. Social hierarchies were still firmly in place, and though service providers considered them all to be Bhutanese, differences had not disappeared. I learned of these "revoked" invitations through service providers, politically engaged lower castes, and from interviews with lower-caste people in Adelaide. It was generally the more politically engaged lower castes that approached service providers to intervene in these situations.

In this example, service providers attempted to transform the Bhutanese into an egalitarian community. They tried to do this by changing the expectations of the lower castes and the behaviours of the higher castes. The lower castes were consistently informed they were part of a Bhutanese community that was welcoming and encompassing. This increased the degree of integration the lower castes expected. The higher castes were informed they had to embrace the lower castes, not as inferiors in a social hierarchy but as equals. The lower castes approached these interactions with elevated hopes of a transformed community, while the higher castes desperately tried to return to a "pure" model without upsetting their benefactors. This balance became difficult to maintain when lower castes were given access to public space or became employees of the service providers.

Generally, the public spaces were used strategically to further political agendas of the higher castes by reflecting the values of the audience. However, the lower castes, due to their partial exclusion, were not as tightly regulated as the higher castes. They used these public spaces in ways the higher castes deemed inappropriate. One lower-caste refugee man used a public forum hosted by the Migrant Resource Centre to discuss the caste system, highlighting the ongoing discrimination perpetrated by the higher castes. In this very public setting, he framed the caste system as being at odds with the egalitarian norms and values of Australia. Further, he questioned the behaviour of the group in Salisbury, suggesting it was attempting to mislead service providers with a performance of community equality. In the days after his talk, the higher-caste Bhutanese in Salisbury spoke of his outburst with disdain (interviews, Adelaide, 2014). They speculated that doing such a thing was a deliberate attempt to undermine their efforts in Australia. They felt he should be punished for such a transgression. Angry participants thought they should attempt to talk to his employer in order to get him fired or, at the very least, ban him from speaking publicly. His public conversation about the internal group dynamics defied not only the image of a cohesive group but also the story of a group bound by the ideals of human rights. Not only was he not behaving as a "good" refugee, he was tarnishing the carefully cultivated image of the group.

This public sharing of behaviours considered by service providers to be "cultural problems" was particularly distressing. Several participants explained to me they have a culture that people in Australia love. This culture includes the way they dress, the food they cook, songs, and their presumably strong sense of community. It was through the sharing of their culture that they gained social recognition in Australia. Cultural problems, on the other hand, threatened not only opportunities for positive recognition but also the relationships they were actively building with their benefactors.

Occasionally, service providers or council employees were approached to intervene on the behalf of lower castes. For example, one Thursday, the meeting of Bhutanese in Salisbury was coming to a close. The guests had departed—only the Bhutanese and their council facilitator remained. The volunteers were preparing to collect teacups and begin the final cleaning, when the facilitator called out for everyone's attention. An incident had been brought to her attention over the weekend: a love marriage had taken place between a higher-caste man and a lower-caste woman. In line with responses I observed in the camp, after the marriage, the man was banned from returning to his home and the bride's family had reluctantly allowed the couple to stay with them. This was in stark contrast to the socially acceptable arrangement after marriage, where the bride joins the groom's home. The son risked not only homelessness but also excommunication (and his family performing premature funeral rites) due to his decision to marry outside his caste. The lower-caste family approached the council member to help find a solution; they did not want to appear to be actively supporting the marriage and thus become subject to further stigmatization. The council worker informed the group, "This kind of behaviour is not acceptable. People here are free to marry whom they want." A high-caste Brahmin priest replied defensively, "This issue does not concern you, this is a community issue and we are capable of fixing it." The council worker countered, "It becomes my issue when people come to me and are upset." The higher-caste participants used the very language promoted in Australia—particularly the notion of community—to distance themselves from incursions by the council staff. Partially due to the unique power dynamic in Salisbury, the facilitator did not push the matter further. Though unsuccessful at persuading the group to defer to her guidance, she was able to maintain her role as a legitimate manager of the Bhutanese. Pursuing the issue further risked the very image both the council and the group had carefully cultivated: a united Bhutanese refugee community. For the higher castes, external intervention into the caste system was the least desirable course of action. As the earlier discussion of the sacred thread ceremony illustrated, they actively tried to maintain their understanding of how society should be organized, even though it had the potential to undermine their status as a deserving community that "fit" into Australia's vision of itself as a nation.

A few of the lower castes had entered into paid employment with refugee or migrant resettlement agencies—particularly the MRC. One woman who was a nurse in Bhutan divided her paid employment in Adelaide between refugee-specific health projects and the MRC. Another lower-caste man, who was a university teacher in Bhutan,[1] was also employed as a community resettlement facilitator through the MRC. One of their tasks was to visit the homes of new arrivals to ensure they were settling in and their basic needs were being met. Due to their position in the caste hierarchy, they were deemed ritually polluting and thus often barred from entering homes. The MRC was alerted to a perceived lack of community cohesion and responded by actively attempting to renegotiate the inner-group power dynamic in Salisbury to foster a more inclusive Bhutanese community.

One such effort was to place the lower-caste woman in the MRC's northern outreach office. Women, as the idealized figure of refugee, were positioned as natural bonding agents, not only holding their community together but mediating between the Bhutanese and broader Australia. Here, the MRC was attempting to "engage and instrumentalize the forces of individuals and groups in the name of the public good" (Rose 1999, 171). The MRC employee was not entirely happy to have to commute to the northern suburbs from her more central suburb. She also seemed unhappy about interacting predominantly with the higher castes. The higher-caste women, in turn, were furious when they learned of her placement at the Salisbury MRC. Higher-caste Bhutanese women utilized the MRC building in Salisbury nearly every day of the week. They held sewing circles and gardening groups. They also used the space to participate in Multicultural Foodies, OzHarvest, and the Penguin Club. They had a high degree of ownership over the space: a few held keys to the building, and a higher-caste Bhutanese woman was one of the centre's

---

1. It is difficult to discern the degree of marginalization lower castes experienced in Bhutan. Based on surveys I conducted in the camps, the lower castes had significantly lower education levels than higher castes. Most participants shared stories that suggested a rigorous caste system with clearly defined social roles. The lower castes who had achieved a high degree of education in Bhutan attributed their educational opportunities to the Royal Government of Bhutan's sponsored study programs.

two employees. The other employee was a refugee from Sudan who allowed the Bhutanese a wide berth to use the space as they saw fit. For the women who used the area on a nearly daily basis, the news that a lower-caste woman was entering their space was an unacceptable development. They complained to me that there was no need for another caseworker in the northern office.

Part of this concern linked to the higher castes' participation in a group called Multicultural Foodies; this was one of the Bhutanese women's favourite groups. Multicultural Foodies encompassed two arenas: learning to grow fruit and vegetables in the Adelaide climate, and expanding the women's repertoire of cooking techniques. Though the Bhutanese women dominated the group in terms of numbers, other refugees from Burundi, Sudan, and Vietnam also participated. Gardening was a favoured pastime of the Bhutanese. Their photos on their mobile phones tended to depict children and grandchildren in their vegetable gardens. They greatly appreciated guidance regarding suitable vegetables for Adelaide's unique growing climate.

In principle, the higher-caste women would not eat with this diverse group of ritually impure people. Yet, aside from one woman, they prepared and ate food together. This woman abstained due to her obligation to fast for her family. Many older women fasted once a week to increase the merit of their family, thus ensuring future success and health. Eating with a diverse group was understood as part of learning to live in Australia. However, preparing food and eating with the lower-caste woman was considered untenable. These higher-caste women wanted to reproduce the structure of community that provided them with power and prestige—they were trying to maintain their collective understanding of a community with clearly defined roles and boundaries. They were willing to subordinate a degree of their ritual purity by eating with non-Hindus but unwilling to concede to the authority of the MRC and abolish the caste hierarchy.

A second activity also illustrated the women's rejection of the value of an egalitarian community. The Penguin Club was a fortnightly meeting held in Salisbury that focused on developing female literacy and promoting confidence in public speaking. These efforts were premised on the notion that some women were disadvantaged and needed additional support to have a "fair go" in Australian society. Many of the higher-caste women who participated were illiterate in both Nepali and in English. There were limited educational opportunities in Bhutan, and though a few participated in the adult literacy classes in the camps, their confidence was still developing. The Penguin Club was coordinated by native Australian English speakers and provided a rare opportunity for the Bhutanese to practise the local English pronunciation. The women who attended were diligent, yet learning a new language was challenging and many expressed embarrassment at their pronunciation. The possibility that the newly appointed lower-caste woman, already possessing these skills, would infringe on these lessons was unacceptable for the higher castes. Worse still was the possibility that caste roles could be inverted, with the lower-caste woman functioning as a teacher for the higher castes.

Rather than framing their concerns in terms of caste and pollution, the higher-caste women questioned the skills and capabilities of the new MRC employee, the lower-caste woman. The lower-caste lady spoke impeccable English. She had worked for an international NGO in Kathmandu rather than living in the camps. She had a vastly different refugee experience from the majority of women living in Salisbury. In Australia, she was employed to host workshops regarding public health and well-being. Recently, she had a hand in compiling recipes from different refugee groups to produce a cookbook. These skills and accomplishments did not elevate her role with the Bhutanese but rather problematized it. This was not because the higher-caste women did not think these skills were useful—quite the contrary. Their efforts with the Penguin Club accentuated this. Indeed, the existing MRC employee in the northern office was a higher-caste Bhutanese who had a similar skill set. Participants spoke glowingly of this woman's efforts, capabilities, and skills. The lower-caste woman held many of the skills the higher-caste women viewed as important and were actively struggling to gain. However, because she was a lower caste, these skills had the effect of further isolating her from the very group the employer wanted her to help transform. One participant informed me she was concerned the lower-caste woman was mentally unstable: "not right" (Brahmin female, age 50s, interview, Salisbury, 2013). This genuine concern stemmed from the perception the lower-caste woman had usurped the social hierarchy in an entirely unacceptable manner. Her capabilities and successes were not celebrated as a demonstration of a downtrodden person succeeding regardless of social obstacles. Rather, her skills accentuated the degradation of the caste hierarchy the higher castes were striving to maintain. Her caste-ordained role was not to hold the employment positions that were considered high-status and in line with "Brahmin" or higher caste-type behaviours.

Having lower castes in sought-after positions with resettlement organizations, particularly when they were tasked with educating the higher castes regarding social norms and values in Australia, was viewed as an affront to the very foundation of a Bhutanese community. As a lower-caste woman, she was expected to serve—not direct— the higher castes. The tension between the performance of an egalitarian community and the desire to maintain a hierarchical social structure became evident at larger events hosted in Salisbury. Often, these events were coordinated by service providers or council staff and were described as community celebrations. The social unease between the groups was perhaps most accentuated when Hindu festivals were hosted in Salisbury.

Community Events

Due to the efforts of the council and the large population of Bhutanese, Salisbury hosted numerous festivals in the Hindu calendar. On these occasions, the lower castes had to weigh whether to enjoy the spectacle of dancing and singing that such an affair promised against the high likelihood of being treated like interlopers. One Saturday morning, I joined the lower castes while they were having their weekly meeting in the Adelaide city centre. That afternoon, the Teej celebration in Salisbury was being held at a local hall the Bhutanese Australia Association of South Australia had secured for the event. The festival was also doubling as a fundraiser for the Bhutanese Youth Radio Station. Service providers, council staff, and the local newspaper were all invited to see "the Bhutanese community celebrate Teej."

Teej is one of the most important festivals in Hinduism. Coinciding with the beginning of the monsoon season, it commemorates a marriage myth centred on the devotion of Goddess Parvati to her husband. This goddess underwent an extended period of fasting in an attempt to beguile her future husband, Lord Shiva. Due to her efforts, she won his favour. This myth highlights the devotion and sacrifice a wife makes for her husband. To celebrate this commitment, married women return to their natal home, reconnecting with family and friends. Historically, due to patrilineal and virilocal marriage practices, this was the only time of the year when a daughter would return to her village. She would only be able to make the journey if her brother accompanied her as an escort. In the maternal village, lavish feasting and then strict fasting occurs as women don red saris and worship Lord Shiva. Singing and dancing convey the heartache women feel for the maternal homes they left behind. This occurs in conjunction with the *pujas* (prayers and offerings) undertaken in the hopes their husbands remain healthy so they will enjoy a long life together.

Though this festival reinforces a male-dominated society (Stirr 2010) through the worship of the husband figure, the women I worked with called Teej "a woman's festival." The event was much anticipated. For most Bhutanese women in Australia, their mothers were either resettled in other countries or still in the camps. A visit to any of these places was difficult due to the formidable cost. Thus, most looked forward to the festivities as a chance to express female solidarity and an opportunity to celebrate the sacrifices they made throughout the year. Leading up to the event in Salisbury, the higher castes spent weeks rehearsing and organizing—the youths perfected Bollywood-inspired dance numbers, with the older women busily writing songs and choreographing new dance steps. The Bhutanese in Salisbury buzzed with excitement, reflected in the invitations they issued to council staff and the local newspaper. For the higher-caste women, it was the jewel in their social calendar.

The group of lower-caste women who met in downtown Adelaide, and particularly the newest arrivals, similarly looked forward to the prospect of celebrating Teej. On the morning of the festival, they proudly displayed the new saris they purchased in anticipation of the event. Yet the members of the lower castes who had been in Australia longer seemed more hesitant. One told me she was only attending because the event was also a fundraiser for the Youth Radio Station and the youth coordinator specifically promised there would not be any caste issues. The youth coordinator explained the highest caste had prepared *momo* (dumplings), so the food would be ritually pure. Additionally, it was packaged individually so the attending lower castes could not touch and pollute the food. These elaborate measures seemed to do little to ease the minds of the potential attendees. If anything, it reminded them of their polluting status in relation to the Bhutanese in Salisbury.

As we walked across downtown Adelaide to the train, we lost close to half of our small group. Those who expressed reluctance to attend disappeared over the four blocks to the train station, quietly turning off onto side streets or ducking into shops. We arrived at the hall in Salisbury to find it already filled with ornately dressed, high-caste women and smartly attired, high-caste men. Every woman's wrist seemed heavy with *chuddas* (bangles), and every man's head was topped with a smart *topi* (hat). In addition to the ornately dressed Bhutanese, several Australians representing various organizations were in attendance. As the dancing and singing began, the lower castes stood in the back of the room and a few remained outside, not wanting to offend the ritually pure higher castes inside. Yet not everyone conformed to this expectation.

During the Teej festival, I spent the majority of time with a higher-caste woman who married a lower-caste man. Due to this marriage, she was considered untouchable. She was not satisfied with this arrangement and consistently tested the boundaries of ritual purity and pollution. Upon our arrival, she immediately navigated me towards the table selling *momo*. We purchased our food, motivated as she explained to me, "to support the youth." Rather than taking our food outside, as would be expected for the lower castes eating around the higher castes, she insisted we begin eating adjacent to the table selling food. As higher castes filtered past the strategically placed table between the door outside and the door to the main theatre area, she offered to share her *momo*. Sharing food and feeding other people are a common occurrence during festivals. The exchange of food tends to reinforce the caste hierarchy: lower castes accept food from higher castes but not vice versa. If food moves down the caste hierarchy, the higher castes should still not touch the recipient. Further, eating in the presence of the lower castes is considered polluting. Some higher castes pretended not to see what was going on, while others stopped when she identified them by name. As my companion cheerfully offered her food, my presence compounded the awkward situation. This woman threatened to undermine the group's cultivated image of an egalitarian community.

The high castes actively strove to minimize outward evidence of the hierarchical implications of the caste system. One of the observable manifestations of the caste system is the refusal to accept food or water from lower castes. While this refusal may not be common knowledge in the broader Australian population, it is one that service providers were acutely aware of. The Bhutanese were also aware of my familiarity with the system. One high-caste man, whom I worked with extensively, very reluctantly accepted a *momo* from her while simultaneously inquiring about my research. She was obviously pleased that a high-caste male ate from a low-caste female in a public arena. Her pleasure, set against his discomfort, articulated the tense negotiations occurring within the group that were consistently approached by service providers and council workers as a community.

Behind the public display of community—the spectacle of dancing in ethnic garbs—internal politics persisted. Though marketed as a community event, this was not an occasion for all Bhutanese. The treatment of the lower castes during the festival illustrated why many decided not to attend: it was clear they were not welcome. They were largely ignored and most did not feel comfortable sitting inside the hall. Those who were bold enough to go inside did not behave in line with caste protocol. In the minds of the higher castes, this further legitimized the need to exclude the lower castes. Several higher castes explained to me that these interactions where physical contact is a possibility are akin to being forced to eat rotten food—the physical effect was profound and sickening. Yet the higher castes eagerly told outsiders that they prided themselves on the happiness they derived from "sitting close," being physically close. Part of this sentiment related to their time in the camps, where the managing institutions stressed community values based on national unity. Further, it mirrored a discourse of humanitarianism that identified community solidarity as an ideal social attribute. In Australia, this sentiment underscored competing notions of community. The higher castes viewed the community they would "sit close" to as those within their strata of caste. Due to a shared nationality, the lower castes, in Australia and in the camps, were lumped into this same community by managing institutions. Service providers attempted to reorganize the group from a hierarchical into an egalitarian community.

Faced with these vastly different notions, it was unsurprising that attempts by service providers only resulted in public performances of unity and acceptance. The lower castes found themselves in an awkward social situation. They were quite aware of the caste-based notions of community, but the well-meaning service providers preserved their hopefulness for an egalitarian transformation.

The exclusion of the lower castes during the Teej festival sent a clear message: the caste system and its associated hierarchy were firmly entrenched and the lower castes had no place in Salisbury. The recently arrived, low-caste woman, who earlier in the day had fawned over her new sari, was systematically excluded. As a low-caste woman (though it would apply to a low-caste male as well), the rules of etiquette did not require the higher castes to welcome her. Yet she obviously thought she would be welcomed because she was now in Australia—a country described by participants as a place where everyone is equal. She fought back tears as we rode the train back to the city. The woman who attempted to share her *momo*, regardless of her small victory, was similarly distraught. She asked me, "Why do we do this to each other?"

Australian values and the service providers who mediated them had changed the expectations of the lower castes. In the camps, exclusion from events hosted by the higher castes was unsurprising. However, the lower castes were not happy with this exclusion, and actively sought forms of community based on alternative religions or heightened ethnic affiliations. In the camps, there was less of a sense that a drastic reform of the caste system was probable. In Australia, service providers consistently emphasized the ideal of Bhutanese unity. They went to great lengths to ensure all Bhutanese were aware of events occurring in Salisbury and, at times, even arranged transportation so those living outside the northern suburbs could participate. Unfortunately, this raised the expectations of the lower castes, who continued to hope these well-meaning invitations were a reflection of a radically different community—one defined by egalitarian rather than hierarchical values. As service providers made efforts to foster a community in line with Australian expectations, most of the lower castes were deciding to look beyond their community and appeal to mainstream social values.

Bhutanese the Australian Way

In Australia, Salisbury emerged as a destination suburb for the higher castes. Several higher castes in the camps, bound for other parts of Australia, stated their desire to move from their initial place of resettlement in order to be closer to their community. A few higher-caste participants who had been resettled in other states in Australia had already made the transition to Salisbury. Originally, some lower castes lived in Salisbury. Due to the degree of exclusion and segregation the lower castes experienced in Salisbury, this group began to explore alternatives. Most moved to different suburbs to avoid interacting with the higher castes. A lower-caste participant clarified, "Many do not go to Salisbury because they don't feel welcome. Sometimes transportation is also an issue, but many just find it easier to avoid" (male, age 30s, interview, Adelaide, 2013). Somewhat ironically, the lower castes had been pushed into more affluent suburbs in closer proximity to universities, with better infrastructure, more educational opportunities, and less violence. Thus, while this settlement pattern was a reincarnation of earlier social norms, in Australia it had unintended social effects.

Woodville is a suburb that has attracted some lower castes. Woodville is far from an affluent suburb. It experienced a period of decline similar to what happened in Salisbury and Elizabeth after the local manufacturing industry collapsed in the 1990s. As a positive outcome of this manufacturing decline, the refugees could afford to purchase homes here. However, unlike Salisbury, it was gradually gentrifying. Woodville enjoyed a more central location, had lower rates of crime than Adelaide's northern suburbs, and had higher rates of tertiary education (ABS 2011, Table 16). Residents holding a bachelor's degree or higher were the fastest growing segment of the population—14 per cent compared to 10 per cent of the broader population of South Australia (ABS 2011, Table 16). Woodville residents are primarily employed as professionals. When I asked lower castes (male and female) if they would ideally live in Salisbury, they responded that Salisbury was violent, "had shootings," and was too far from the universities where their children studied (interviews, Adelaide, 2012 and 2013). Though nearly all Bhutanese under the age of thirty were enrolled in some sort of educational program, many of the highest castes were getting skilled training from the technical and vocational school, TAFE, located in Salisbury. A Certificate

III in Home and Community Care appeared to be the most widespread training course for the higher castes in Salisbury. This is an entry-level qualification that commands a salary roughly half of the average wage within South Australia (ABS 2011). The lower castes, living in more central suburbs, were attending university with the hope of becoming nurses and engineers, and two young women were pursuing a desire to become doctors. This suggests the lowest castes were trying to establish their social status with respect to the wider Australian population, as a high degree of education did not enhance their position within a caste-conscious community.

While in the camps, conversions to different religions were viewed as one means of opting out of the caste system. In Australia, many lower castes were actively shunning the many visual signifiers of caste. Clothing and related adornments of the body are highly significant, particularly for women. One informant suggested one could tell virtually everything about a person's caste based on their dress (Brahmin female, age 30s, interview, Nepal, 2012). Women of the highest castes are visually striking, with elaborate (and hierarchically specific) *tikkas*, heavy green necklaces bound with beaten gold, colourful saris, *chuddas*, nose piercings, and waist-length hair. They are quite literally wearing their caste on their sleeves—though this is not obvious to the Australian public. To the untrained eye, some Bhutanese simply looked more exotic than others.

Several of the lower-caste women had removed markers that would signify to the trained eye their caste and had adopted a Western appearance. One young, lower-caste lady told me, "[I] lost my nose piercing down the sink, my mother was very mad but I didn't want to put it back in" (Kami female, age early 20s, interview, Adelaide, 2012). Another frankly explained she had removed her piercing because she did not like its caste associations (Gazmere female, age 40s, interview, Adelaide, 2012). Though not all of the low-caste women had switched to wearing pants, they were doing so with greater frequency than the highest castes. As the lower castes shed these exotic trappings, the image of Bhutanese-ness became even more high-caste centred. Inadvertently, this reinforced the caste hierarchy by naturalizing the higher castes as the representatives of exotic, cultural, visible Bhutanese-ness in Australia. However, it moved the lower castes into alternative spaces within Australia that are not necessarily defined by their ethnic community. Because their population was smaller, and they were effectively rejected from the main group, the lower castes were striking out on their own. A point of discussion for the lower castes during their Saturday meetings was their progress making friends. They frequently lamented this was difficult, often due to language difficulties but also the impression Australians were "without the time" to talk to new people. Regardless, making friends with the broader population was deemed important, necessary, and achievable. For the higher castes, their engagement with the broader population was mediated through a well-defined multicultural framework: ethnic groups interacted with ethnic groups via service providers. For the lower castes, because they were outside the group, the emphasis shifted towards individuals interacting with individuals, rather than communities interacting with communities. Their exclusion from Salisbury drastically changed their experience of resettlement, effectively accelerating the need to make connections beyond their group's boundaries. The internal power dynamic of the group shifted as the lower castes accessed alternative opportunities outside institutionally mediated relationships with broader Australia.

Articulations of expected Bhutanese-ness in Australia privileged and constrained people in different ways. As the high-caste Bhutanese became wedded to the performance of a bound ethnic community, they risked becoming perpetually peripheral subjects versus central participants. Some seemed to be aware of this danger. A Brahmin man explained to me, "We need a strong community to keep the culture, the traditions, but to find a job and improve our economy, our community cannot help. All of them are beggars" (unknown age, interview, Salisbury, 2013). The lower castes, though consistently orientated towards "their" community by service providers, were actively pursuing alternatives. They were able to tap into broader notions of community in an attempt to transform their social roles. However, for the Bhutanese, the roles available to them in Australia ran counter to the hopes they accumulated in the camps. To bridge the gap between their hopes and their experiences, some drew more vigorously on the notion of the caste system. Others, I would say many, did hope to move beyond the parameters of their "refugee/Bhutanese community" and achieve greater integration into Australia. They attempted to transform their role as incapable victims. In doing so, they constructed themselves as vital to the society of their new hosts. However, the necessity of this construction and reconstruction highlights the precarious acceptance refugees experience in Australia.

The Bhutanese began to form opinions regarding the role of refugees in Australia while still in refugee camps. An earlier refugee settlement policy in Australia strove to accept those with the most settlement potential (Jupp 1995). This is still the expectation of most migrants to Australia. The Bhutanese faced a lack of foreseeable durable solutions in the camps, which is the primary reason they were being resettled. However, the humanitarian approach to refugees in Australia appears to mirror the resettlement priorities set forth by the UNHCR (2011, 37): survivors of torture/violence; those with medical needs; women, girls, and youths at risk; and families that need to be reunited should be given first priority. In 2016, Australia allocated five thousand places for its "special humanitarian program" (Elibritt 2016, 7). These places are set aside for humanitarian entrants—that is, people who are subject to substantial discrimination amounting to a gross violation of human rights in their home country, such as "women at risk" (McAdam 2013, 438).

During the resettlement screening process, people had very little control over their destination. With the UNHCR as their proxy, refugees viewed nations as selecting refugees based on attributes they considered desirable. The United States was understood as a country that sought refugees who could work; Australia was viewed as a country that wanted refugees to care for (traumatized, disabled, or otherwise vulnerable). This understanding was not necessarily an accurate reflection of the intention of Australia's policies and priorities but was widespread. It was expressed in the following quote by a Bhutanese man resettled in Australia: "At first, I wanted to go to America. There it is easy to find the job. Because I was a victim of torture the UNHCR decided I needed to go to Australia" (Brahmin male, age 40s, interview, Salisbury, 2012). A woman in Salisbury, who wanted to bring her mother from the camps to Australia, explained the perceived differences between the countries:

> They [UNHCR] want my mum go to the USA because my brother is there. I want her to come to Australia with me. She has this eye problem and Australia is better if you are sick. What could she do in America? Here

she will have some support (Magar female, age 30s, interview, Salisbury, 2013).

Australia does not provide support for a refugee that exceeds the provisions for natural-born citizens. Rather, refugees made comparisons to the United States because the Australian welfare system is, overall, more robust. Crucially, these two quotes illustrate the impact of the Australia government's recent emphasis on accepting vulnerable refugees coming from UNHCR-run camps. While noble, these understandings can "facilitate and solidify processes of exclusion and marginalization in different contexts of displacement" (Fiddian-Qasmiyeh 2010, 64). It illustrated the way refugees interpreted the selective criteria for admission: refugees were expected to be vulnerable and in need of additional support.

Dependency is not only expected but, in some regards, deliberately sought through a focus on those who are physically diseased or suffering from debilitating trauma (Fassin 2005; Fassin and Rechtman 2009; Ticktin 2006). Once resettled, the attributes that facilitated their admission based on compassion have ongoing consequences. Arriving as presumably incapable communities, refugees are perceived as being in acute need of their benevolent host's assistance (Ong 2003; McKinnon 2008; Colic-Peisker 2005). In the Netherlands, the way the welfare state helps refugees "is such that it transforms active participants into passive dependants on the state" (Ghorashi 2005, 195). In turn, their passivity and dependence mark them as a social problem requiring charitable assistance rather than members of a broader society entitled to support. Further, in Finland, robust social welfare arguably obscures other forms of social marginalization and exclusion during refugees' experiences of resettlement (Valtonen 2004). Robust financial support fetters criticisms of broader practices of hiring discrimination and a general underrepresentation of refugees in the political arena (Valtonen 2004). It also illuminates the creation of two subjects: "On the one hand, the subject is passive and pathetic, the one who suffers. On the other, the subject is active, the one who identifies suffering and knows how to act" (Englund 2006, 32).

For refugees, "helping" effectively masks broader practices of social exclusion—both in terms of a strict migration policy and the degree of social integration refugees can hope to achieve. Changing imaginings of refugees raises barriers against integration into their new societies. Though the context of Australia exhibits unique attributes, it also demonstrates that the way refugees are helped can similarly confine them to the role of dependent subcitizens. In response to the narrow scope of participation, the Bhutanese created roles for themselves not defined by their status as recipients of charity but as contributors to the nation.

Crafting Effective Narratives

A politically charged, male-centric origin story dominated the public sphere. Men were fiercely possessive of their democratic story. During a family interview, I asked the youngest daughter what she thought about the refugee camps she had grown up in. "The camps, they are just horrible. They are—" Her father interrupted, "The older people have knowledge the younger people can't explain, our real problems, the census, and our government's policy" (family interview, Salisbury, 2012). This interaction articulated the intergenerational tensions that many participants felt resettlement accentuated. It also illustrated that certain stories were considered more valid than others and certain people more credible narrators. If I was speaking with a married man about democracy or political advocacy and requested his wife participate, I was met with a rebuttal. The following response, by a man in his mid-thirties, was typical: "Oh, my wife doesn't understand these things. Women don't have time for politics. They have to worry about the house and those things. They are very busy" (Brahmin male, age 30s, interview, Adelaide, 2013). Nonetheless, the women I spoke with had well-developed, nuanced understandings of democratic values and the group's history. Yet the powerful men in Adelaide only wanted me to speak with other powerful men. While this may reflect an older domestic–public dichotomy, it was also a reaction to the limited roles available to male refugees. Both men and women perceived women as having greater social flexibility in Australia while a central social role, motherhood, was still available to them. Men, on the other hand, recognized their ideal social role as the families' economic providers was becoming unrealistic. In response, they consistently tried to present themselves as political activists in exile.

Powerful Bhutanese men promoted a story based on democracy and egalitarian aspirations. It exemplified "how they valorised themselves in order to negotiate their predicament" (Gatrell 2013, 283). The following story was shared at a public event hosted by STTARS in Adelaide. It was held in a public space adjacent to the main shopping centre in Salisbury. The coordinator and the Bhutanese participant went to great efforts to set up displays that chronicled their journey to Australia, as well as their cultural and national heritage. There were "Nepali" swords sitting alongside "Bhutanese" darts, and ethnic costumes were similarly intermeshed on a folding table covered with woven fabric. It was a windy day, so the Bhutanese were kept busy ensuring carefully selected photographs, photocopies of legal documents, and drawings from the camps did not blow away. Though unique to the speaker, the following narrative was illustrative of the themes the men strove to promote:

> *Today I am going to talk about how we reached here in Australia from Bhutan. Bhutan was ruled by the Wangchuck dynasty since 1907 and was without a written Constitution, Bill of Rights, or Rule of Law until 2008. We used to live in peace and harmony until the fourth King enacted a policy of one people/one nation. He forced the north Bhutanese culture, language, and customs on the south Bhutanese with threat of the death penalty. Ones who spoke against the government and appealed to human rights faced the death penalty. Many people were tortured and imprisoned during this time. We tried to appeal to the King to lift these terrible and treacherous things. Instead of hearing our appeal, they employed the army and forced the south Bhutanese into exile in Nepal. We fought for many years for democracy and human rights in Bhutan. Now, there is a so-called democracy but we are still not enjoying. After living for so many years in the refugee camps, we were forced to accept the option of resettlement. With that option we are now in Australia. We are very thankful to be with you all (Brahmin male, age 40s, public event, 2013).*

This story touches on the key themes male participants consistently shared both with me and in public spaces: that the Bhutanese are refugees because of their political values and actions undertaken to promote democracy.

Women, on the other hand, tended to minimize the causes of their exile when speaking publicly and instead focused on the assistance they received. The following speech was shared at an International Women's Day luncheon hosted by the Migrant Research Centre in downtown Adelaide:

My life history...I was born in Bhutan—it is a very lovely, mountainous country. I left and the most miserable days of my life were in the refugee camp in Nepal. Slowly, organizations like the UNHCR and Caritas came. I spent nearly eighteen years in the refugee camp before I got opted to come to Australia through the UNHCR. In Australia, I faced many difficulties including language and the transportation systems. I began to attend language courses at TAFE in Salisbury...I started volunteering for MRC, in all I spent eighteen months as a volunteer and luckily there was a job for Bhutanese settlement officers. I love my current job because I like helping the people here. Many thanks to the MRC and the Australian government for empowering me (Chhetri female, age 30s, public event, 2013).

While women would discuss the complex, politically charged events that led to and (perhaps) maintained their exile in private, these were rarely shared publicly. For men, the publicly shared backstory centred on political action became very important. It relayed a version of the conditions that forced the refugees into exile: namely a tyrannical king imposing on a minority group and an extended period of time in refugee camps. This story suggested the Bhutanese did not willingly forego their nation: first they were forced out, and only after many years did they accept resettlement. Their camp experience demonstrated they were good refugees and was proof they came to Australia through the correct channels. This narrative stressed the significance of reassuring people they arrived in Australia through a particular process—affirming the sovereignty of Australia's borders. In turn, how refugees arrived became evidence of the moral aptitude of the group—a group deserving of support. Both of these talks, given at public events, underscore that the Bhutanese were self-consciously highlighting a narrative that affirmed the strongly gendered expectations of the host community.

This narrative was more than mirroring expectations; the refugees were actively renegotiating what refugee status entailed. In Australia, their status as refugees required that the Bhutanese represent the humanitarian heart of Australia: they were admitted due to compassion rather than capabilities. The stories that circulated were carefully selected to confirm suffering and cast the community as genuine or deserving refugees. These tropes assured the audience (and by extension, the nation) of their own goodness in accepting the Bhutanese for resettlement. The story that was told was the story the Bhutanese thought the audience wanted to hear.

These expectations and responses illustrated that "being a refugee also naturally suggested, even demanded, certain kinds of social conduct and moral stances, while precluding others" (Malkki 1996, 381). The Bhutanese men were precluded from some forms of social conduct and responded by taking a strategic moral stance. In response to stifling expectations, the Bhutanese reincarnated an earlier understanding of refugees based on political skills as a barometer for future social contributions. The publicly promoted version of experiences justified their presence in Australia while subtly rejecting a language of victimization—asserting an alternative identity based on political activism. In this new understanding, they were not "helped because they are helpless" (Stein 1981, 185) but supported because they were skilled. Rather than their political experience marking them as crippled by trauma, it centres on their abilities. They worked within what they perceived as a framework of expectations that assumed—and perhaps demanded—that a male refugee had experienced trauma.

Crucially, though, they are renegotiating the implications of such a story. Rather than traumatic experiences resulting in them being perpetually victims, their traumatic experiences are a reason they should be viewed as equals. During resettlement, they are actively breaking down the binary between "refugee as incapable" and "citizen as capable" by underscoring that the very experience of trauma has made them more capable of contributing to the nation of Australia. A man explained:

> *I have three layers of identity. First, I am a Bhutanese citizen, second I am a refugee, and third I am Australian. I have been three citizens in my life. Citizen in Bhutan then I became a refugee/human rights advocate*

and now I am okay. I will get citizenship next year. It has been like this, morning, day, and night (Brahmin male, age 50s, interview, Adelaide, 2021).

The men's efforts to present themselves as freedom fighters pushed against stereotypes of refugees as passive, incapable recipients. By highlighting a version of events in which they were political freedom fighters, the Bhutanese countered assumptions of both statelessness and lack of institutional strength. This provided a path for them to shift social margins ever so slightly closer to the mainstream. As the earlier quote illustrates, the status of refugee and human rights advocate conjoins to lead to a final identity: Australian citizen. Problematically, the limited pathways available to the men in Australia constrained the sharing of alternative stories and experiences. Because an effective narrative was created, it became subject to careful regulation in order to maintain its operative potential. Men, in particular, had to manoeuvre between being simultaneously "an object of care and a source of anxiety" (Feldman and Ticktin 2010, 6).

The Bhutanese reframed their experiences to orientate themselves away from the constraining role of apolitical, dependent victims. They did this by adopting a moral interpretation of their origins as refugees. This refashioning of the causes of their exile was akin to the process of mytho-history making Malkki (1995a) observed in the Burundi refugee camps. In that context, refugees created a historic trajectory that gave moral credence to their present situation. Malkki's (1995a) analysis focused on the public performance of "refugee-ness" and the formulaic narratives that buttressed it. While this approach underscored the coherence of the Hutus' mytho-history, analyzing experiences considered outside the realm of public performance also merited consideration. Values and ideals masked in public performances provided illuminating insights regarding hierarchies within a specific group, as well as the broader hierarchy of humanitarian values they negotiated.

During fieldwork, competing versions of the public narrative emerged. Service providers deemed the caste system problematic, both in the camps and Australia. The persistence of this system threatened not only notions of community social relations but egalitarian aspirations. During the events that led to exile from Bhutan, some of the higher castes protested against the government because clothing and accessories were central ways of relaying information about the Hindu-based caste structure—ensuring that ritual purity was maintained. It was difficult to avoid a polluting person if they were not clearly distinguished. By extension, because Brahmins, and particularly Vishnu Brahmins, saw their task as the protectors of Hinduism, these changes threatened to undermine the broader social institution of Hinduism. Democratic reforms could, in the minds of the higher castes, help them maintain a social status that was coming under threat. A Brahmin priest explained, "Personally, I used to perform ceremonies and give the red *tikka*, our worship was too strong so the Drukpa [government] banned it and that is the reason I am here" (male, age 60s, interview, Adelaide, 2012). Several participants revealed they left Bhutan because they worried they would be forced to consume beef or convert to Buddhism. These were overtly religious concerns.

While freedom of religion is considered a basic human right, and fleeing due to religious persecution is enshrined as a legal way to claim protection as a refugee, this motivation was not the most readily repeated story. Motivations were reconfigured from preserving the caste system to promoting democracy. This distancing from alternative (though related) motivations may have been in response to the number of interventions in the camps and, in Australia, criticisms of the caste system. The Bhutanese were well aware of the contingent nature of their deserving status, the thin line between acceptance and rejection. Distancing themselves from versions that could appear at odds with the moral framework of Australia required them to relegate some of their most deeply held convictions and beliefs to a subsidiary role.

The lower castes also sought democratic reforms and were some of the original leaders in Bhutan. Similarly, most participants asserted they wanted to protect a Hindu-based culture that was perceived as being under attack by the Bhutanese government. However, they saw democracy as a chance to transform the caste system by enhancing the social and political power of the lower castes. Democracy was viewed as the same means to different ends. One low-caste participant in Australia identified the caste system in Bhutan as the key reason their movement was not successful:

The prime reason that we were not more successful in Bhutan was because the caste divides...These issues could have all been resolved if people could have come together but it was so fragmented (Gazmere male, age 40s, interview, Adelaide, 2012).

These complex motivations and competing understandings of their exile were not frequently shared publicly. These "cracks" threatened to undermine the cultivated image of a "Bhutanese refugee community" and, by extension, their inclusion in the public space of Australia. Their narrative, on the other hand, contained the ideals they wanted to be associated with. The carefully crafted and regulated story had been sanitized of competing notions of community and disputes over the image of a democratic world. Instead, it reflected values refugees felt they must perform.

The reinterpretation of the events leading to, and the experience of, exile provided a way for men to maintain a functional position for themselves in the face of rapidly changing gender roles and shrinking possibilities in Australia. The moral interpretation of exile presented one of the few pathways towards establishing themselves as respected equals within their new country. Espousing the values of democracy and equality provided the only legitimate way to counter a constraining recognition based on victimization.

The Bhutanese, as the previous sections argued, were not satisfied with being the recipients of compassion and actively sought to transform their social role. Sharing their narratives was a significant and important means of achieving that. Yet these opportunities orbited around their experiences as refugees. The credible narrative hinged on past actions. This both created and rewarded "a limited version of what it means to be human" (Ticktin 2006, 34). For the Bhutanese in Australia, to be accepted as a human was to be perpetually a refugee. Though the Bhutanese were able to use the limited parameters of accepted "refugee-ness" in a strategic manner, there was also a strong push against a role defined by the past.

Contributing as Australians

The spaces provided to refugees to speak publicly often had a clear, predefined agenda. Generally, refugees spoke at events designed to raise funds or awareness and centred on refugees sharing their experience of flight or exile. In Australia, there were limited platforms to showcase contemporary actions—past experiences continued to define what it was to be a refugee, and "traditional" manifestations of culture (dancing, clothing, distinct foods) were used to define a community. The Bhutanese leaders were not satisfied with this. Public spaces were sought out and creatively employed to illustrate they were more than their past experiences but contemporary contributors. For example, in 2012 the Migrant Resource Centre hosted an Ethnic Leaders Forum to discuss the redevelopment of a suburban community centre. After government representatives shared the proposed plan, the floor was opened up to the audience for input regarding facilities different groups thought were necessary. The president of BAASA, sitting with several Bhutanese men, raised his hand to speak:

> *I am not sure if you are aware, but* BAASA *has organized to host the 2013 interstate soccer match. As you know, we have organized several programs that have been very successful. We would like to do even more. The top Bhutanese teams from all over Australia will be playing here in Adelaide. We have been organizing all of this. If there was some priority to soccer fields, that would be good. We are looking for support and sponsorship, it is going to be a very big event. There will be hundreds of people coming. We will also have several important performances with Bhutanese dancers and singers that will allow for the wider community to develop an appreciation of other cultures (Brahmin male, age late 20s, public event, Adelaide, 2012).*

His statement is surprising because the community centre slated for redevelopment was in a suburb where few Bhutanese lived. Further, the project was not going to be completed until the end of 2015; the upcoming soccer tournament had little relation to the redevelopment.

The participation by the Bhutanese man suggested more than a desire to partake in the planning process of a suburb in Adelaide. It was a strategic political presentation. Though the Bhutanese were asking for additional support, the message was clearly that they had undertaken this endeavour on their own accord. They were capable organizers and planners. Rather than passive recipients of charitable gestures, they reframed themselves as equals in a new country. They presented themselves as more than just "good refugees"—they were actively creating a clear role as contributors to Australia. They aspired to be viewed "as people *with capacities*, in short, people 'like us'" (Harrell-Bond and Voutira 2007, 282). Further, they were reflecting values that had the potential to move them into the mainstream. Sport and fair play strongly underpinned the "social, national and cultural landscape" in Australia (Zakus et al. 2009, 994). Organizing an interstate soccer tournament was an attempt to connect with the national character, reflecting their similarity to the broader community. The need to seek out and transform unexpected, and perhaps inappropriate, spaces demonstrated the constraining social framework the Bhutanese inhabit. As a community of refugees, they were wedded to a label that rewarded a narrow cannon of experiences and behaviours. "Refugee" became an inescapable essence that allowed them entry into Australia while curtailing their ability to move beyond the role of recipient.

Many participants viewed formal citizenship as the final step on a long journey—the official transformation from refugee to citizen. Gaining citizenship in Australia required a four-year residency period and the passing of an English-based test. In 2013, the group in Salisbury devoted considerable time preparing everyone who reached the residency requirement for the test. They coordinated with the Department of Immigration and Border Protection to have a "special" day during the 2013 Refugee Week when 185 Bhutanese took their citizenship oaths. One excited man explained before the event,

> *There are many migrants here in Australia, but the organization is doing this just for the Bhutanese...we have been here only for four years and already they are giving us citizenship...They [the Department of Immigration and Border Protection] are going to do all of this for us, they are going to do the ceremony, refreshments, everything. They'll call the head authorities, maybe in Canberra, and the media—the media will be there (Brahmin male, age 40s, interview, Salisbury, 2013).*

The perception of special treatment and talk of an expedited process were conceived as a means of verifying their status as deserving refugees. The perception that their democratic ideals and good behaviour allowed them to gain citizenship status sooner than other migrant groups validated the interpretation of their journey to Australia. Rather than passive recipients of social welfare, traumatized and incapable, they had become vital members of Australia.

Citizenship was widely viewed as the realization of political voice. A man in his sixties, looking forward to citizenship, explained, "I will have a say with my voting rights. If there is something or some point in the Constitution that I have a question about then I can ask it" (Chhetri male, interview, Adelaide, 2012). Many parents expressed the hope that their children would hold elected roles in government. One former activist from Bhutan suggested, "To come to Australia, a new country, and see my children elected would be wonderful" (Chhetri male, age late 40s, interview, Adelaide, 2013). Again, this goal mirrored the expectations they encountered in Australia: refugees will not be able to contribute, but if their resettlement is successful, their children will. Only through their children, who were theoretically not as firmly wedded to a refugee identity, could they hope to transform into full members of Australian society.

Most adult participants, even after gaining citizenship, remained bound to the refugee label. This understanding illustrates the limits of formal citizenship and the constraints that contemporary approaches to refugees presented; "although equality of citizenship rights can be taken as a starting point, this legal equality does not necessarily lead to equality of respect, resources, opportunities or welfare" (Valtonen 2004, 75). While the Bhutanese were quick to describe themselves as special refugees, as refugee citizens they were still defined as not quite equal to the Australian citizen.

Not Quite Equals

Despite attempts to incorporate their support in Australia into a preexisting framework of familial obligation, or by creating roles for themselves as contributors, social support and acceptance in Australia were still perceived as discretionary. Though the Bhutanese strove to present themselves as equals to the broader population of Australia, they found themselves in a paradoxical relationship. While it was reiterated that all humans are equal, their status as recipients of compassion demanded humility—positioning them as slightly unequal to those with the power to give (Fassin 2012). There was uncertainty that refugees had a *right* to be in Australia and a strong impression they had been resettled due to a charitable gesture. As a charitable gesture, "it constitutes a relationship of dependency, not of equivalence" (Calhoun 2010, 35). While the motivations for accepting and caring for vulnerable refugees may be framed as purely altruistic, some Bhutanese still found it degrading. One man explained, "We can say we are Australian now, but we are all still beggars" (Brahmin male, age 60s, interview, Adelaide, 2012). Though most participants imagined their children would shed the refugee label and its associated relationship as beneficiaries, some expressed apprehension that refugee status would follow their children: "I am happy that the government wants to take care of us, but I worry that our children, even if they have a high level of education, will not be able to get jobs" (Chhetri male, age 40s, interview, Adelaide, 2012). The awareness of this unequal relationship impacted on the everyday life of the Bhutanese, producing a fixation on exhibiting good behaviour.

The Bhutanese had a hyper-awareness of the potential negative consequences of mundane transgressions. For example, using pedestrian crossings was viewed as a marker of a well-behaved refugee. One woman, who had been in Australia for several years, explained to me that they came from a place where these things were not important, but in Australia "they represent the law" (Gazmere female, age late 30s, interview, Adelaide, 2012). Many worried the transgression of crossing the street outside the acceptable parameters could potentially jeopardize their status as deserving refugees. Behaviour such as urinating in public, not caring for others in the community, and not being respectful of service providers meshed together to form what were considered dangerous markers of social deviance.

Participants of all ages frequently worried that poor behaviour could form a legitimate basis for complete exclusion from Australia—a possible return to exile in camps. While Australia is a signatory to the 1951 UN convention that protects asylum seekers and refugees from refoulement, in 2014 amendments to Australia's Migration Act stated, "Australian officials may remove a person...from Australia without considering whether or not they are at risk for refoulement" (McAdam and Chong 2019, 10). Australia seems to be actively discriminating against "refugees based on their mode of arrival" (McAdam and Chong 2019, 26), and the policy, as far as I am aware, has not returned people to UNHCR-run camps. However, the Bhutanese experienced exile before and were well aware that a shaky distinction between asylum seeker and refugee in policy and popular rhetoric could impact on them. One lady, who recently gained citizenship, said, "For a long time I was very scared the local people would throw us out" (Brahmin female, age 20s, interview, Adelaide, 2012). Another refugee, a young man, explained, "We are only recently born here, we've not grown up here yet so we cannot do silly things" (Chhetri male, late teens, interview, Adelaide, 2013).

These concerns were not isolated to people who had arrived days, weeks, or even months earlier. Some of them had arrived in Australia in 2008, and still others held Australian citizenship. The persistence of these concerns articulates the refugees' sense of precarious acceptance. The perceived risk of losing their position as deserving refugees functioned to regulate their everyday choices and experiences. Because resettlement was framed as a charitable gesture, rather than a right, refugees remained beholden to the nation long after they had shed the official status of refugees and became citizens. The Bhutanese were experiencing "a politics of compassion that emphasizes benevolence over justice, standards of charity over those of obligation—or that ultimately protects and encourages a limited and limiting notion of humanity" (Ticktin 2006, 42). In turn, what appeared to be compassionate and benevolent ultimately favoured the interests of the nation over the rights of people.

In January 2014, Australia fulfilled its quota of Bhutanese refugees. For the Bhutanese in Australia, there were still family members in the camps. These included wives who were left behind when polygamous families had to divorce and the children who stayed with their mothers. There was the possibility they could be reunited through a family reunification visa that was available to all refugees in Australia. The Bhutanese viewed this visa as being allocated based on which groups of refugees were considered most deserving and best behaved. Reuniting families was considered a privilege reserved for good refugees.

Though the Bhutanese built a very effective narrative regarding their political skills, they did not rely on this skill set to sway the Department of Immigration and Border Protection to reunite their families. There were no political marches to demand resettlement rights, or hunger strikes undertaken to increase the visa quota of Bhutanese. These political protest techniques were relatively common in the camps. Rather, in Australia, the Bhutanese emphasized their good behaviour, their community values, and the vulnerability of those they wanted to bring to Australia. *Demanding* changes to the resettlement system challenged the very attributes (helpless, passive, and malleable) that marked them as deserving. While democratic values were promoted as a crucial Australian attribute, it was the value of human suffering they engaged with to further their cause. The humanitarian gesture thus had a role in "helping to reproduce the geopolitical order because it reduce[d] pressures that might have demanded its transformation" (Barnett 2005, 733). As some of the "most vulnerable" fell through the cracks of resettlement, the possibility of demanding resettlement rights became even more remote. The democratic ideals the refugees placed at the centre of understanding their exile were claimed to be incompatible with the reality in Bhutan. But they were compatible with the ideals of Australia. However, engagement in the political discussion, particularly regarding who should be allowed to enter Australia, remains a slim possibility. The Bhutanese found few means to directly agitate for political, economic, or social transformation. There has been some tinkering in the margins but few demands. In turn, the broader paradoxes within humanitarian ideals become evident. Australia's desire to help the helpless has the consequence of precluding them from political participation.

Conclusion

The Australian service providers were aware of the complicated social dynamics within the group but desired a united community—this came into conflict with the internal organization of the Bhutanese. Caste was a central way the Bhutanese, and particularly the highest castes, understood the world. For many in Australia, and Salisbury in particular, resettlement represented a long-awaited opportunity to practise their culture by living in line with ritualized, hierarchical forms of purity and pollution. Yet there were pre-existing expectations in Australia regarding the behaviour of communities. Ethnic communities served as the foundation of multiculturalism in Australia, and new groups had to actively try to conform to this understanding. Communities were expected to be coherent, to care for each other, and to fit into specific models of ethnic behaviours. A caste-based community suggested suffering—unacceptable inequality that demanded reform.

Multiculturalism, humanitarianism, and the caste system (with their many differences) are moral systems that attempt to organize people. By extension, it is through these norms that societies are regulated. Unequal power relationships within the expected community become the site of intervention, obscuring the broader power imbalance between benefactors and beneficiaries. Thus, refugees must remain alert to the expectations of their benefactors, carefully balancing their cultural values within a predefined parameter of acceptable behaviours. Similar to their experience in the camps, refugees have to conceal certain constructions of themselves while highlighting others. There are complex motivations behind the resulting performances. These performances are an imposition but also a political strategy. An image of community unity is a political asset for powerful Bhutanese leaders, particularly when they can promote an effective narrative that counters the constraining understandings of what it is to be a refugee.

Conclusion

HUMANITARIAN GESTURES

IN SEPTEMBER 2015, images of a Syrian toddler's corpse, washed ashore in Turkey following his family's failed attempt to reach Europe, evoked both sympathy and outrage around the globe. The photograph sparked a public outcry that became a powerful call for action. The prime minister of Australia, Tony Abbott, who consistently held firm that the aggregate number of refugees resettled in Australia must not be increased, announced the acceptance of an additional twelve thousand Syrian refugees:

> *This is a generous response to the current emergency...the response best reflects Australia's proud history as a country with a generous heart. The focus will be those people; women, children, and families who have sought temporary shelter...I do want to stress, women, children, and families—the most vulnerable of all...it is important that we don't bring anyone from this troubled region who might ultimately be a problem for the Australian community (Abbott 2015).*

Abbott repeated further that Syrian asylum seekers who try to reach Australia independently would not be welcome. The position makes clear the paradoxes with the contemporary humanitarian paradigm: a select few "will be recognized through humanitarian reason rather than the right to asylum" (Fassin 2012, 157). Compassion and moral outrage seemed to bring about a political response that promised to help the most vulnerable of refugees. The "generous gesture" appeared to be a sound victory for laudable humanitarian ideals. Yet this victory had an unsettling foundation: actions premised on righteousness obscured the fact that refugees have a *human* and *legal right* to seek asylum. By replacing rights with compassion, humanitarian discourse provides an additional space for the sovereign performance of statecraft by creating contingencies on the basis of who is deserving of admission.

By November 2015, compassion and moral outrage formed a decidedly different response. A series of attacks in Paris left over one hundred people dead. A suspected attacker was found with a forged Syrian passport. Solidarity quickly gave way to suspicion. Syrian refugees no longer were part of a "global community" entitled to support but rather a "different humanity" providing a conduit through which terrorists could move with impunity. Further complicating this division was that the majority of attackers were French or Belgian nationals, descendants of earlier generations of immigrants. This raised broader concerns this "different humanity" could never "be transformed into nationals of the country" (Arendt 1951, 301). The threat from outside appeared to have found a way inside. The earlier generous gesture to resettle refugees seemed increasingly precarious as the moral concern for citizens trumped the need to alleviate the suffering of refugees. These responses highlight the tensions between emancipation and domination, community and diversity. These responses seem to be fundamentally at odds, yet they articulate the duality and contradictions within the contemporary global order.

When we look at the migration trajectory of the Bhutanese, it becomes clear that both international and domestic forms of humanitarianism are transformational projects focused on the behaviour of victims rather than the conditions that cause suffering. We tend to think of humanitarian endeavours as primarily done for the recipients—alleviating suffering, improving their lives. However, it is more complex than that. Humanitarian gestures not only help and transform people in need; they provide absolution for the societies in the position to make such gestures. Just as the caste system was embedded into the refugees' community, humanitarian ideals and actions are embedded in the system of nations in multiple, complex, and at times contradictory ways. Humanitarian ideals are at once a noble aspiration and a fickle agent for the inequality embedded in the nation-based order of our world. They both help normalize national boundaries while striving to protect those that fall outside of those protections. These are complementary approaches to ordering our world. Supporting camps and accepting a few refugees for resettlement are framed as gestures based on the compassion of those in wealthier, safer, more powerful countries. Humanitarian ideals speak to our greatest aspirations, but ultimately they have not radically transformed the conditions that give rise to displacement.

Humanitarianism as Governance
Across the world, the processes of building new nations, solidifying old boundaries, or creating a democratic order are pushing people out. These refugees are devoid of the potent symbolism that defined earlier Cold War era refugees as capable, political actors defying Communist ideologies. Instead, they become embedded ideologically in a newly possible, universal understanding of a globally connected humanity. Rather than political actors in exile, refugees have become a "miserable sea of humanity" (Malkki 1996) in need of care, protection, and—crucially—guidance. A humanitarian response premised on compassion (Feldman 2012) has become the central means of alleviating the suffering masses. Refugees articulate both the grand aspirations and the desperate shortcomings of a new humanitarian system of global relationships.

The international community responds to humanitarian emergencies in a very specific way. These responses are

> *nestled in discourses of compassion, responsibility, and care, which, in turn, are attached to claims that the "international community" has obligations to its weakest members. This international humanitarian order has all of the elements of governance (Barnett 2012, 486).*

"Refugee" has become a social category of exception that affords people recognition in a broader humanitarian hierarchy, but only as the perennially "weakest members." This hierarchy is strictly regulated by excluding some causes while simultaneously *producing public representation of the human beings being defended (e.g. showing them as victims rather than combatants and by displaying their condition in terms of suffering rather than the geopolitical situation) (Fassin 2007, 501).*

This process of tailoring causes and representatives of these causes is intrinsically linked to governance. The "miserable sea of humanity" must be formed into acceptable humanitarian subjects. Refugees become transformed into victims who need compassion rather than people who have internationally recognized rights.

Despite aspirations of global equality, national borders are still the crucial means of organizing our world. Camps become surrogates for nations, almost quasi-nations governing people and regulating borders. A key "lesson" refugees learned in the camps was the significance of national boundaries as a means of organizing social interactions. However, this system of organization frequently finds itself at odds with the ideal of radical equality—the equivalence of all human life. Thus, the nation and its populace find a "constant need to redefine the threshold in life that distinguishes and separates what is inside from what is outside" (Agamben 1998, 131). The act of setting a threshold and redefining it accentuates that it is the sovereign "who decides on the exception" (Schmitt 1985, 5). Humanitarian organizations, funded by countries that can afford to support them, have a central role in a system of global governance.

Humanitarian governance flourishes by focusing on suffering, and "helping" provides a moral authority that is difficult to dispute. When the exception becomes the most desperate, vulnerable, and traumatized, this process becomes powerfully cathartic for the hosting nation and the nations that financially support the camps. It is here that humanitarianism finds its greatest strength; "it fugaciously and illusorily bridges the contradictions of our world, and makes the intolerableness of its injustices somewhat bearable" (Fassin 2012, xii). Malkki (1995b) has argued that the nation has been naturalized as a means of ordering the world. Humanitarian configurations and humanitarian ideals have a role in maintaining the contemporary global order of things.

In the context of refugees, a system of governance has emerged that is analogous to Ferguson's (1990) critiques of development in Lesotho: a vision of what needs to be done is created and then projected onto a situation. The resulting "solution" to the problem is increasingly removed from conversations regarding broader economic, geopolitical, or social arrangements (Ferguson 1990). How refugees became and remain refugees no longer seems to matter. The solution is increasingly focused on reforming or redeveloping refugees, rather than demanding broader political action or a reconfiguration of global relationships. In short, the politics behind refugee generation and containment is curiously absent. Instead of focusing on broader political solutions, the domestic institutions of refugees are evaluated and found to need fixing. As an ideology and a system of governance, humanitarianism provides a means for analyzing the tension between helping the weak and controlling the many. Humanitarian governance is focused on two key processes: transforming the refugee and absolving the nation.

Humanitarian Governance as Transformation
Refugees are helped in very specific ways that go beyond government by containment. While refugee camps have been critiqued as spaces that warehouse people (Smith 2004), the bustle of activity to improve refugees has transformed camps into "workshops." The new society birthed in camps is composed of good refugees. Good refugees willingly accept criticism and remould themselves to hold the "right" values. These transformative endeavours underscore how "legitimate refugee life is often defined in the seemingly contradictory nexus of apolitical victim and the improving subject" (Feldman 2014, 245). As Malkki (1995a), Feldman (2014), and Gabiam (2012) observe, camps play a central role in creating, moulding, and transforming the lives of refugees. This is not a quirk of multiple and diverse refugee camps but illustrative of a broader system of governance. Humanitarianism is both an ideology and a path for action. This morphing, globally ordering humanitarianism has moved beyond the bare roots of saving lives towards improving the human condition through moral transformation.

Activities in the camps seek to transform and rebuild people who are aspiring towards emancipation. Camps become a spot for potential development and improvement—a space where international values can bloom. Exploitative gender relations are transformed, children are made aware of their rights, and equality permeates the idealized refugee communities. These lofty goals are undertaken with the conviction they will change refugees' lives for the better. Limited funds and the threat of compassion fatigue provide a strong incentive for refugees to conform to these improving efforts. Some of these efforts may herald transformations, but they all have unintended consequences. Barnett (2011) observes that, in a humanitarian framework, power is frequently exercised over those it hopes to emancipate. Thus, as humanitarianism strives to help liberate, it also regulates. The pathway to liberation is framed in virtuous terms and the parameters of that virtuous path are created by the organizations that govern. Crucially, the focus on transforming refugees does not challenge the state system and may, in fact, create a more stable, legitimate state system (Barnett 2005, 733). Camps become a means not of generating equals but of crafting amendable subjects. The transformations of humanitarian subjects bring about the promised emancipation, but it is ultimately a limited kind of freedom.

Becoming morally acceptable requires refugees to transform, but once transformed their status is still precarious. Funding camps or accepting people for resettlement is presented by nations as a generous gesture, giving the impression to recipients that such gestures can be easily revoked. Once resettled, refugees find their status still conditional and contingent on the good graces of their hosts. Though legally accepted, economic and social marginality persists. In this context, refugees must conform to the clear parameters of acceptable refugee behaviours. These frameworks provide little freedom to directly critique experiences, become active participants in resettlement, or pose broader questions regarding global arrangements. Rewarding select behaviours while problematizing others functions to silence competing values, thus maintaining the status quo.

The status refugees acquire is precarious and requires continual image management as refugees try to maintain their relatively privileged status (compared to asylum seekers). This illustrates Arendt's (1951) argument that if people become reduced to an externally defined category, such as the suffering refugee, they lose not only their freedom but also risk becoming a "specimen" rather than a political or social actor. The experiences of the Bhutanese illustrate the complex strands of power that intertwine to limit the scope of people who "deserve compassion."

This is not to suggest that humanitarian action is completely oppressive, or that all refugees are oppressed. Refugees are not passive in these situations. Rather, they actively work within the framework available to them. They do this by questioning the resources provided and presenting a specific image of community, or through constructing a finely honed refugee narrative. Feldman (2012, 164) observed a similar process in the Palestinian context, arguing that though humanitarian action can fetter people's ability to act, it can simultaneously provide a framework for people to make claims that are not solely humanitarian. Despite these actions that "tinker" in the margins, most demands move down the humanitarian hierarchy. The desire to reform or transform refugees illuminates that humanitarian governance provides few mechanisms to balance the power imbalance between governed and governor.

Humanitarian Governance as Absolution

Camps are an integral part of a system of international governance built on the sovereignty of nation-states. They embody the dual model of humanitarian governance: well-sealed boundaries between the poor countries that house refugees and the wealthy countries that fund the camps. Yet keeping millions in camps indefinitely is morally unacceptable—a constant reminder of global inequality. Faced with protracted exiles and little chance of resettling, effective management and development programs become a path towards absolution, placating the troubled conscience of an "international community."

The numerous efforts undertaken to improve the camps and create virtuous, humanitarian subjects all normalize the camps. A well-run camp, helping refugees improve themselves in line with international values, inadvertently makes the spaces morally acceptable. These efforts, funded by a largely unseen international community, suggest that refugees are included, however marginally, in the broader global system. In turn, this helps legitimize refugees' containment, further institutionalizing refugee camps. As the raw suffering and horror subsides, and the images fade from the news, the thread linking the global community with the camps becomes tenuous. Yet this tenuous link is still significant, providing a "salutary power for us because by saving lives, it saves something of our idea of ourselves...it also relieves the burden of this unequal world order" (Fassin 2012, 252). Humanitarian gestures not only help and transform people in need; they provide absolution for the societies in the position to make such gestures.

Wealthier countries that strictly regulate their borders provide the financial and ideological support for the camp systems. Australia, as with many wealthy countries, strives to be compassionate while still being tough on border security. In Australia, the focus on accepting the most vulnerable is used to exclude the many while also regulating the few who are allowed entry. These competing notions create a moral tension between hostility and hospitality that refugees must negotiate once they are resettled. This constraining framework, based on "helping vulnerable refugees," can become an effective way to justify broader forms of exclusion. Regulation is achieved and normalized when it is framed as an attempt to support the refugees who are most deserving: refugees provide the moral absolution the wealthy countries desperately desire.

Humanitarian ideals quickly confront a situation in which not everyone can be helped. A hierarchy emerges, ranking those considered deserving against those who fall outside the accepted parameters. Refugees, even after resettlement, find themselves perpetual humanitarian subjects rather than capable, potential citizens. This is a constraining, liminal role with little space for manoeuvring. Refugees are still beholden to those with the power to offer sanctuary. Increasingly, the role of the few elite refugees who are granted the opportunity to resettle is to provide absolution for wealthier countries that strictly regulate their boarders.

Fassin (2012, 10) implores anthropologists to produce critically engaged ethnographies examining the impact of the humanitarian value system on the everyday experience of its subjects. He calls for the "study of the production, circulation and appropriation of norms, values, sensibilities, and emotions in contemporary societies...concerns we could easily take for granted, sometimes even viewing as moral progress." Bornstein and Redfield (2011, 25) similarly observe, "There are relatively few in-depth ethnographic and historical accounts of humanitarian organizations, cosmologies, and encounters." In response to these concerns, this book examines the cosmologies and encounters the Bhutanese have with humanitarian sentiments. In doing so, it provides a historically situated and critically engaged account that illustrates the relationship humanitarianism has with larger configurations of politics and power. It traces the movement of moral sentiments across the obviously humanitarian camps and the less obviously humanitarian spaces afforded to refugees in Australia.

Humanitarianism is not fundamentally wrong, nor do humanitarian efforts, either in the camps or through social welfare in Australia, need to stop. Humanitarian efforts and sentiments can, indeed, provide crucial, life-saving benefits. However, humanitarian ideology and action, while claiming to have the well-being of the most vulnerable at its core, can lead to complicity. Focusing on relieving immediate suffering can have the effect of leaving larger questions regarding global inequalities and broader injustices unasked. Ticktin (2011, 223), referring to asylum seekers in France, argues for a transformed perception of refugees and asylum seekers that is not based on finding the exceptional but recognizing the "current unequal access to the means of existence."

Closing Thoughts

There were roughly six thousand Bhutanese refugees still living in camps at the beginning of 2022 and these people are not going to be resettled. Those who remain have diverse motivations that are only briefly touched on in this book. Some still hope for repatriation; others want to die in the country they are familiar with. The UNHCR and its associated organizations phased out involvement as the resettlement program concluded in 2016. Those who remain are still refugees but no longer living under international stewardship. What it is to be a Bhutanese refugee in this future space that is not a camp but also not quite part of Nepal could provide fruitful lines of academic inquiry, as Banki and Phillips (2017) have started to explore.

Other avenues of research could explore the relationship the Bhutanese in Australia have with legal citizenship. While my research was conducted when they were gaining citizenship in great numbers, the impact of official belonging merits greater examination. Theoretically, citizenship is the final shift away from humanitarian subject towards full political actor. It would be interesting to see if and how official citizenship has an impact on their perceived position in Australian society, specifically in relation to their perceived ability to make a meaningful contribution to the nation.

More generally, this research did not focus on NGOs, though some organizations and employees became inadvertent participants. There has been fruitful research in this arena by Bornstein (2012), yet it remains an area that could produce compelling insights regarding the motivations and understandings of those in a position to help. NGOs both in the camps and in Australia actively work to transform the lives of refugees, materially and ideologically. Examining the relationship between NGOs and those they help could produce interesting horizontal and vertical ethnographic accounts. While this kind of dual focus requires considerable time, it could reveal tensions and ambiguities this book has overlooked.

Humanitarian action can, and does, provide important benefits to the societies it serves. It is an ideology and a path of action that increasingly focuses on governing by transforming the groups it strives to help. Given its power and potential, a robust interrogation of both its intended and unintended consequences is necessary. As Feldman and Ticktin (2010, 25) argue, "We may not be able to do without it…but we have to remain uneasy with its deployment." Careful scrutiny must be directed at the lived experiences of those who are the beneficiaries of these benevolent gestures. Humanitarian situations will undoubtedly continue to arise throughout the twenty-first century. Focusing on a select few deemed morally legitimate "can work to preclude responses to and by the suffering" (Ticktin 2006, 222). We do not need to abolish humanitarian ideals or stop aspiring to stem suffering and inequality. Rather, because "humanitarian governance is not just a small corner of international order—it is a growing and increasingly prominent dominion" (Barnett 2012, 487), we must remain mindful of the limits such an approach can impose.

In concluding, I return to the situation in Europe. By November 2015, over 870,000 people entered Europe—the vast majority seeking asylum from Syria, Afghanistan, and Iraq (IOM 2015b). After the attacks on Paris, refugees morphed into a group no longer eligible for compassion with the same startling speed that months earlier had elevated them into the sphere of "deserving." The suspected actions of a few people "proved" these refugees are fundamentally different from members of a global community. The earlier humanitarian response became a kind of mirage, an "illusion of equality of human beings in the face of misfortune" (Fassin 2005, 397). Governors in the United States (Healy and Bosman 2015) moved rapidly to free themselves from any obligation to resettle refugees, and countries in Europe began to tighten their borders (Erlanger 2015). As refugees become a threat to the safety of citizens, their potential admission into Europe and the United States may become barred in good conscience. This shifting landscape reveals the dangers of searching for the most deserving, the most vulnerable, and the morally legitimate recipients of compassion. A scope narrowed by even the most humane values (compassion, vulnerability) still necessarily precludes many who deserve recognition and redress. These decisions have very real consequences regarding who is allowed access to a safer world and who is barred from entry.

References

Abbott, T 2013, *Address to the 2013 federal coalition campaign launch*, transcript, viewed 11 November 2015, http://www.liberal.org.au/latest-news/2013/08/25/tony-abbott-address-2013-federal-coalition-campaign-launch.

Abbott, T 2015, *Australia to take in 12 000 Syrian refugees*, online video, viewed 26 November 2015, http://www.abc.net.au/news/2015-09-06/abbott-vows-australia-will-help-syria-refugees/6753220.

Abu-Lughod, L 1991, "Writing against culture." In RG Fox (ed.), *Recapturing anthropology: Working in the present*, School of American Research Press, Santa Fe, pp. 137–162.

Adams, V 1996, *Tigers of the snow and other virtual Sherpas: An ethnography of Himalayan encounters*, Princeton University Press, Princeton.

Adamson, FB and Tsourapas, G 2019, "Migration diplomacy in world politics," *International Studies Perspectives*, vol. 20, no. 2, pp. 113–128.

Adamson, FB and Tsourapas, G 2020, "The migration state in the global south: Nationalizing, developmental, and neoliberal models of migration management," *International Migration Review*, vol. 54, no. 3, pp. 853–882.

Adelman, H 1998, "Why refugee warriors are threats," *Journal of Conflict Studies*, vol. 18, no. 1, pp. 49–69.

Agamben, G 1998, *Homo sacer: Sovereign power and bare life*, trans. D Heller-Roazen, Stanford University Press, Stanford.

Agamben, G 2005, *State of exception*, trans. K Attell, University of Chicago Press, Chicago.

Ager, A and Strang, A 2008, "Understanding integration: A conceptual framework," *Journal of Refugee Studies*, vol. 21, no. 2, pp. 166–191.

Agier, M 2010, *Managing the undesirables: Refugee camps and humanitarian government*, Polity Press, Cambridge.

Amnesty International 1992a, *Bhutan: Human rights violations against the Nepali-speaking population in the south*, ASA 14/04/92, Amnesty International, London.

Amnesty International 1992b, "First AI visit to Bhutan," *Amnesty International Newsletter*, vol. XXII, no. 4, p. 1.

Ansari, M 2012, *A Shangri-la economy: Exploring Buddhist Bhutan*, Universal-Publishers, Boca Raton.

Arendt, H 1951, *The origins of totalitarianism*, Harcourt, New York.

Arendt, H 2013 [1958], *The human condition*, 3rd edition, University of Chicago Press, Chicago.

Aris, M 1994, "Introduction." In M Aris and M Hutt (eds.), *Bhutan: Aspects of culture and development*, Kiscadale Publications, Gartmore, pp. 7–25.

Aris, M 2005, *The raven crown: The origins of Buddhist monarchy in Bhutan*, Serindia Publications, Chicago.

Ashton, J 1996, "Nepal: All things to all people," *Refugees Magazine*, Issue 104, United High Commissioner for Refugees, Geneva.

Attorney-General's Department 2015, *Crime prevention and community safety grants*, Government of South Australia, Adelaide.

Australia Bureau of Statistics (ABS) 1911, *Census of the Commonwealth of Australia*, Cat. no. 2122.0, Australia Bureau of Statistics, Canberra.

Australia Bureau of Statistics (ABS) 2011, *Population for local government areas*, Cat. no. 1376.0, Australia Bureau of Statistics, Canberra.

Australia Bureau of Statistics (ABS) 2014, *National regional profile*, Cat. no. 1379.0.55.002, Australia Bureau of Statistics, Canberra.

Australia Refugee Association (ARA) 2014, *Annual report 2013–2014*, Australia Refugee Association, Underdale.

Australian Government Department of Home Affairs 2020, *Life in Australia: Our values and principles*, Commonwealth of Australia, Belconnen.

Bakewell, O 2003, "Community services in refugee aid programs: The challenges of expectations, principles, and practice," *PRAXIS: The Fletcher Journal of International Development*, vol. 18, pp. 5–18.

Banki, S 2008a, "Resettlement of the Bhutanese from Nepal: The durable solution discourse." In H Adelman (ed.), *Protracted displacement in Asia: No place to call home*, Routledge, London, pp. 27–56.

Banki, S 2008b, *Bhutanese refugees in Nepal: Anticipating the impact of resettlement*, Australia Research Council, Australia National University, Canberra.

Banki, S and Phillips, N 2017, "Leaving in droves from the orange groves: The Nepali-Bhutanese refugee experience and the diminishing of dignity." In E Sieh and J McGregor (eds.), *Human dignity*, Palgrave Macmillan, London, pp. 335–352.

Barnett, L 2002, "Global governance and the evolution of the international refugee regime," *International Journal of Refugee Law*, vol. 14, no. 2/3, pp. 238–262.

Barnett, M 2001, "Humanitarianism with a sovereign face: UNHCR in the global undertow," *International Migration Review*, vol. 35, no. 1, pp. 244–277.

Barnett, M 2005, "Humanitarianism transformed," *Perspectives on Politics*, Issue 4, pp. 723–740.

Barnett, M 2018, "Human rights, humanitarianism, and the practices of humanity," *International Theory*, vol. 10, no. 3, pp. 314–349.

Barnett, M and Weiss, TG 2008, *Humanitarianism in question: Politics, power, ethics*, Cornell University Press, Long Grove.

Barnett, MN 2011, *Empire of humanity: A history of humanitarianism*, Cornell University Press, Ithaca.

Barnett, MN 2012, "International paternalism and humanitarian governance," *Global Constitutionalism*, vol. 1, no. 3, pp. 485–521.

Barnett, MN 2013, "Humanitarian governance," *Annual Review of Political Science*, vol. 16, pp. 379–398.

Barth, F 1969, *Ethnic groups and boundaries: The social organization of culture difference*, Waveland Press, Long Grove.

Basnet, S, Johnston, L and Longhurst, R 2020, "Embodying 'accidental ethnography': Staying overnight with former Bhutanese refugees in Aotearoa New Zealand," *Social & Cultural Geography*, vol. 21, no. 2, pp. 207–221.

Baumann, G 1992, "Ritual implicates 'others': Rereading Durkheim in a plural society." In D de Coppet (ed.), *Understanding rituals*, Routledge, New York, pp. 97–116.

Becker, HS 1967, "Whose side are we on?" *Social Problems*, vol. 14, no. 3, pp. 239–247.

Bennett, E, Newman, L, Burnside, JW, Phatarfod, B, Thomas, RM, Moodie, AR and Moore, MJ 2017, "Ending our shame: Call for a fundamental reconsideration of Australian refugee policy," *The Lancet*, vol. 390, no. 10094, p. 552.

Bennett, L 1983, *Dangerous wives and sacred sisters*, Mandala Book Point, Kathmandu.

Berg, B 2004, *Qualitative research for the social sciences*, Allyn and Bacon, Boston.

Berreman, GD 1962, *Behind many masks: Ethnography and impression management in a Himalayan village*, Society for Applied Anthropology Monograph Number 4, Cornell University, Ithaca.

Besky, S 2013, *The Darjeeling distinction: Labor and justice on fair-trade tea plantations in India*, University of California Press, Berkeley.

Birendra, KC and Nagata, S 2006, "Refugee impact on collective management of forest resources: A case study of Bhutanese refugees in Nepal's eastern terai region," *Journal of Forest Research*, vol. 11, no. 5, pp. 305–311.

Bissell, WC 2007, "Casting a long shadow: Colonial categories, cultural identities, and cosmopolitan spaces in globalizing Africa," *African Identities*, vol. 5, no. 2, pp. 181–197.

Bornstein, E 2012, *Disquieting gifts: Humanitarianism in New Delhi*, Stanford University Press, Stanford.

Bornstein, E and Redfield, P (eds.) 2011, *Forces of compassion: Humanitarianism between ethics and politics*, School for Advanced Research Press, Santa Fe.

Brohman, J 1995, "Universalism, eurocentrism, and ideological bias in development studies: From modernisation to neoliberalism," *Third World Quarterly*, vol. 16, no. 1, pp. 121–162.

Brown, M 2010, "Changing authentic identities: Evidence from Taiwan and China," *Journal of the Royal Anthropological Institute*, vol. 16, no. 3, pp. 459–479.

Brown, T 2001, "Improving quality and attainment in refugee schools: The case of the Bhutanese refugees in Nepal." In J Crisp, C Talbot, D Cipollone and B Daiana (eds.), *Learning for a future: Refugee education in developing countries*, United Nations High Commissioner for Refugees (UNHCR), Geneva, pp. 109–161.

Brown, W 2009, *Regulating aversion: Tolerance in the age of identity and empire*, Princeton University Press, Princeton.

Bryman, A 1988, *Quantity and quality in social research*, Routledge, New York.

Buchowski, M 2006, "The specter of orientalism in Europe: From exotic other to stigmatized brother," *Anthropological Quarterly*, vol. 79, no. 3, pp. 463–482.

Burkert, C 1997, "Defining Maithil identity: Who is in charge?" In DN Gellner, J Pfaff-Czarnecka and J Whelpton (eds.), *Nationalism and ethnicity in a Hindu kingdom: The politics of culture in contemporary Nepal*, Overseas Publishers Association, Harwood, pp. 241–273.

Calhoun, C 2010, "The idea of emergency: humanitarian action and global (dis)order." In D Fassin and M Pandolfi (eds.), *Contemporary states of emergency: The politics of military and humanitarian interventions*, Zone Books, New York, pp. 29–59.

Caritas-Nepal 2021, *Caritas-Nepal annual report 2020-2021*, Caritas-Nepal, Kathmandu.

Castles, S 1988, *Demographic change and the development of a multicultural society in Australia*, Occasional Paper 1, Centre for Multicultural Studies, University of Wollongong, Wollongong.

Castles, S 1992, "The Australian model of immigration and multiculturalism: Is it applicable to Europe?" *International Migration Review*, vol. 26, no. 2, pp. 549–567.

Castles, S 1995, "How nation-states respond to immigration and ethnic diversity," *Journal of Ethnic and Migration Studies*, vol. 21, no. 3, pp. 293–308.

Central Bureau of Statistics 2012, *Nepal living standard survey volume II 2010–2011*, National Planning Commission Secretariat, Government of Nepal, Kathmandu.

Central Bureau of Statistics 2014, *Population atlas of Nepal*, National Planning Commission Secretariat, Government of Nepal, Kathmandu.

Cevik, S and Sevin, E 2017, "A quest for soft power: Turkey and the Syrian refugee crisis," *Journal of Communication Management*, vol. 21, no. 4, pp. 399–410.

Charities Aid Foundation 2021, *World giving index 2021: A global pandemic special report*, viewed 31 March 2022, https://www.cafonline.org/docs/default-source/about-us-research/cafworldgivingindex2021_report_web2_100621.pdf.

Chatterjee, P 2012, "After subaltern studies," *Economic and Political Weekly*, vol. XLVII, no. 35, pp. 44–49.

Chimni, BS 1998, "The geopolitics of refugee studies: A view from the South," *Journal of Refugee Studies*, vol. 11, no. 4, pp. 350–374.

Chimni, BS 2000, "Globalization, humanitarianism and the erosion of refugee protection," *Journal of Refugee Studies*, vol. 13, no. 3, pp. 243–263.

Chimni, BS 2009, "Birth of a discipline: From refugee to forced migration studies," *The Journal of Refugee Studies*, vol. 22, no. 1, pp. 11–29.

City of Salisbury 2014, *City of Salisbury submission inquiry into the local government and cost shifting*, City of Salisbury, Salisbury.

Colic-Peisker, V 2005, "'At least you're the right colour': Identity and social inclusion of Bosnian refugees in Australia," *Journal of Ethnic and Migration Studies*, vol. 31, no. 44, pp. 615–638.

Colic-Peisker, V 2009, "Visibility, settlement success and life satisfaction in three refugee communities in Australia," *Ethnicities*, vol. 9, no. 2, pp. 175–199.

Colic-Peisker, V and Farquharson, K 2011, "Introduction: A new era in Australian multiculturalism? The need for critical interrogation," *Journal of Intercultural Studies*, vol. 32, no. 6, pp. 579–586.

Colic-Peisker, V and Tilbury, F 2006, "Integration into the Australian labour market: The experience of three 'visibly different' groups of recently arrived refugees," *International Migration*, vol. 45, no. 1, pp. 59–85.

Collins, J 2013, "Rethinking Australian immigration and immigrant settlement policy," *Journal of Intercultural Studies*, vol. 34, no. 2, pp. 160–177.

Colson, E 2007, "Linkage methodology: No man is an island," *Journal of Refugee Studies*, vol. 20, no. 2, pp. 320–333.

Commonwealth of Australia 2020, *Australian citizenship: Our common bond*, Commonwealth of Australia, Belconnen.

Cornwall, A and Brock, K 2005, "What do buzzwords do for development policy? A critical look at 'participation,' 'empowerment' and 'poverty reduction,'" *Third World Quarterly*, vol. 26, no. 7, pp. 1043–1060.

Cottle, S and Nolan, D 2007, "Global humanitarianism and the changing media-field: Everyone was dying for footage," *Journalism Studies*, vol. 8, no. 6, pp. 862–878.

Dasgupta, A 1999, "Ethnic problems and movements for autonomy in Darjeeling," *Social Scientist*, vol. 21, no. 11/12, pp. 47–68.

Dauvergne, C 1999a, "Amorality and humanitarianism in immigration law," *Osgood Hall Law Journal*, vol. 37, no. 3, pp. 597–623.

Dauvergne, C 1999b, "Confronting chaos: Migration law responds to images of disorder," *Res publica*, vol. 5, no. 1, pp. 21–43.

Dauvergne, C 2005, *Humanitarianism, identity, and nation: Migration laws of Australia and Canada*, University of British Columbia Press, Vancouver.

de Smet, S, Rousseau, C, Deruddere, N, Kevers, R, Spaas, C, Missotten, L, Stalpaert, C and De Haene, L 2021, "A contextualized perspective on research participation in collaborative refugee research: A multi-site exploration of relational dynamics in collaborative research," *Cultural Diversity and Ethnic Minority Psychology*, Advanced Online Publication, https://doi.org/10.1037/cdp0000458.

Deakin, A 1901, "Immigration Restriction Bill," *Debates*, House of Representatives, Commonwealth of Australia, Melbourne, p. 4805.

Denzin, NK 2012, "Triangulation 2.0," *Journal of Mixed Methods Research*, vol. 6, no. 2, pp. 80–88.

Department of Employment 2016, *Small area labour markets March quarter 2016*, Labour Market Research and Analysis Branch, Labour Market Survey Group, The Australian Government, Canberra.

Department of Immigration and Border Protection 2013, *Information paper December 2013*, The Australian Government, Canberra.

Department of Immigration and Border Protection 2014, *Settler by country of birth (Settlement by state)*, The Australian Government, Canberra.

Department of Immigration and Border Protection 2015, *Australian citizenship pledge*, The Australian Government, Canberra.

Department of Social Services 2012, *Beginning a life in Australia: Welcome to Australia*, The Australian Government, Canberra.

Department of Social Services 2014, *Fact sheet—Australia's multicultural policy*, The Australian Government, Canberra.

Department of Social Services 2015, *Fact sheet 98—settlement services for refugees*, National Communications Branch, The Australian Government, Canberra.

Dhakal, DNS and Strawn, C 1994, *Bhutan: A movement in exile*, Nataraj Books, Springfield.

Donini, A 2010, "The far side: The meta functions of humanitarianism in a globalised world," *Disasters*, vol. 34, suppl. 2, pp. S220–S237.

Douglas, M 2003, *Purity and danger: An analysis of concepts of pollution and taboo*, 3rd edition, Routledge Classics, New York.

Douzinas, C 2007, "The many faces of humanitarianism," *Parrhesia*, vol. 2, pp. 1–28.

Duffield, M 2007, *Development, security and unending war: Governing the world of peoples*, Polity Press, Cambridge.

Duffield, M 2010, "The liberal way of development and the development–security impasse: Exploring the global life–chance divide," *Security Dialogue*, vol. 41, no. 1, pp. 53–76.

Duffy, M 2005, "Performing identity within a multicultural framework," *Social & Cultural Geography*, vol. 6, no. 5, pp. 677–692.

Durkheim, E 2008 [1912], *The elementary forms of religious life*, trans. JW Swain, Dover Publications, Mineola.

Edwards, A 2005, "Human rights, refugees, and the right 'to enjoy' asylum," *International Journal of Refugee Law*, vol. 17, no. 2, pp. 293–330.

Elibritt, K 2016, "Refugee resettlement to Australia: What are the facts," Research Paper Series, Parliament of Australia, Canberra.

Englund, H 2006, *Prisoners of freedom: Human rights and the African poor*, University of California Press, Berkeley.

Erlanger, S 2015, "Paris attacks force European Union to act on border controls," *The New York Times*, 20 November, viewed 23 November 2015, https://www.nytimes.com/2015/11/21/world/europe/paris-attacks-force-european-union-to-act-on-border-controls.html.

Escobar, A 1995, *Encountering development: The making and unmaking of the third world*, Princeton University Press, Princeton.

Essential Media Communications 2015, *Treatment of asylum seekers*, viewed 11 November 2015, http://www.essentialvision.com.au/treatment-of-asylum-seekers-2.

Evans, R 2010a, "The perils of being a borderland people: On the Lhotshampas of Bhutan," *Contemporary South Asia*, vol. 18, no. 1, pp. 25–42.

Evans, R 2010b, "Cultural expression as political rhetoric: Young Bhutanese refugees' collective action for social change," *Contemporary South Asia*, vol. 18, no. 3, pp. 305–317.

Eversole, R 2010, "Remaking participation: Challenges for community development practice," *Community Development Journal*, vol. 47, no. 1, pp. 29–41.

Fadlalla, AH 2009, "Contested borders of (in)humanity: Sudanese refugees and the mediation of suffering and subaltern visibilities," *Urban Anthropology and Studies of Cultural Systems and World Economic Development*, vol. 38, no. 1, pp. 79–120.

Faier, L 2009, *Intimate encounters: Filipina women and the remaking of rural Japan*, University of California Press, Berkeley.

Falzon, MA 2009, "Introduction: Multi-sited ethnography: Theory, praxis, and locality in contemporary research." In MA Falzon (ed.), *Multi-sited ethnography: Theory, praxis, and locality in contemporary research*, Ashgate Publishing Limited, Surrey, pp. 165-179.

Fanjoy, M, Ingraham, H, Khoury, C and Osman, A 2005, *Expectations & experiences of resettlement: Sudanese refugees' perspectives on their journeys from Egypt to Australia, Canada, and the United States*, Forced Migration and Refugee Studies Program, The American University, Cairo.

Fassin, D 2005, "Compassion and repression: The moral economy of immigration policies in France," *Cultural Anthropology*, vol. 20, no. 3, pp. 362-387.

Fassin, D 2007, "Humanitarianism as a politics of life," *Public Culture*, vol. 19, no. 3, pp. 499-520.

Fassin, D 2008, "Beyond good and evil? Questioning the anthropological discomfort with morals," *Anthropological Theory*, vol. 8, no. 4, pp. 333-344.

Fassin, D 2012, *Humanitarian reason: A moral history of the present*, University of California Press, Berkeley.

Fassin, D and Rechtman, R 2009, *The empire of trauma: An inquiry into the condition of victimhood*, Princeton University Press, Princeton.

Fassin, D and Stoczkowski, W 2008, "Should anthropology be moral? A debate," *Anthropological Theory*, vol. 8, no. 4, pp. 331-332.

Feldman, I 2007, "Difficult distinctions: Refugee law, humanitarian practice, and political identification in Gaza," *Cultural Anthropology*, vol. 22, no. 1, pp. 129-169.

Feldman, I 2008, "Refusing invisibility: Documentation and memorialization in Palestinian refugee claims," *Journal of Refugee Studies*, vol. 21, no. 4, pp. 498-516.

Feldman, I 2009, "Gaza's humanitarianism problem," *Journal of Palestine Studies*, vol. 38, no. 3, pp. 22-37.

Feldman, I 2012, "The humanitarian condition: Palestinian refugees and the politics of living," *Humanity: An International Journal of Human Rights, Humanitarianism, and Development*, vol. 3, no. 2, pp. 155-172.

Feldman, I 2014, "What is a camp? Legitimate refugee lives in spaces of long-term displacement," *Geoforum*, vol. 66, pp. 244-252.

Feldman, I and Ticktin, M 2010, *In the name of humanity: The government of threat and care*, Duke University Press, Durham.

Ferguson, J 1990, *The anti-politics machine: "Development," "depoliticization," and bureaucratic power in Lesotho*, Cambridge University Press, Cambridge.

Fiddian-Qasmiyeh, E 2010, "'Ideal' refugee women and gender equality mainstreaming in the Sahrawi refugee camps: 'Good practice' for whom?" *Refugee Survey Quarterly*, vol. 29, no. 2, pp. 64-84.

Fiddian-Qasmiyeh, E 2011, "The pragmatics of performance: Putting 'faith' in aid in the Sahrawi refugee camps," *Journal of Refugee Studies*, vol. 24, no. 3, pp. 533–547.

Fiddian-Qasmiyeh, E 2014, *The ideal refugees: Gender, Islam, and the Sahrawi politics of survival*, Syracuse University Press, Syracuse.

Fisher, WF 1997, "Doing good? The politics and antipolitics of NGO practices," *Annual Review of Anthropology*, vol. 26, pp. 439–464.

FitzGerald, DS 2019, *Refuge beyond reach: How rich democracies repel asylum seekers*, Oxford University Press, New York.

Fitzgerald, J 2007, *Big white lie: Chinese Australians in white Australia*, University of New South Wales Press, Sydney.

Flick, U 2004, "Triangulation in qualitative research." In U Flick, E von Kardoff and I Steinke (eds.), *A companion to qualitative research*, Sage Publications, Thousand Oaks, pp. 178–183.

Foster, RJ 1991, "Making national cultures in the global ecumene," *Annual Review of Anthropology*, vol. 20, pp. 235–260.

Fox, F 2001, "New humanitarianism: Does it provide a moral banner for the 21st century?" *Disasters*, vol. 25, no. 4, pp. 275–289.

Fox, RG 1989, *Gandhian utopia: Experiments with culture*, Beacon Press, Boston.

Francisco-Menchavez, V 2020, "Researching Queenila and living in-between: Multi-sited ethnography, migrant epistemology and transnational families," *Migration and Development*, vol. 9, no. 1, pp. 56–73.

Friedman, J 1992, *Empowerment: The politics of alternative development*, Blackwell Publishers, Cambridge.

Gabiam, N 2012, "When 'humanitarianism' becomes 'development': The politics of international aid in Syria's Palestinian refugee camps," *American Anthropologist*, vol. 114, no. 1, pp. 95–107.

Galbally, F 1978, *Review of post arrival programs and services for migrants*, Migrant Services and Programs Canberra, Australian Government Publishing Service, Canberra.

Gallagher, A 2002, "Trafficking, smuggling and human rights: Tricks and treaties," *Forced Migration Review*, vol. 12, no. 25, pp. 8–36.

Gallenkamp, M 2011, "The history of institutional change in the Kingdom of Bhutan: A tale of vision, resolve, and power," Working Paper Number 6, Heidelberg Papers in South Asian and Comparative Politics, South Asian Institute, Heidelberg University, Heidelberg.

Gatrell, P 2013, *The making of the modern refugee*, Oxford University Press, Oxford.

Gellner, DN 2007, "Caste, ethnicity and inequality in Nepal," *Economic and Political Weekly*, vol. 42, no. 20, pp. 1823–1828.

Ghelli, T 2014, "UNHCR pilots new biometrics system in Malawi refugee camp," *Making a Difference*, 22 January, United Nations High Commissioner for Refugees, Geneva.

Ghorashi, H 2005, "Agents of change or passive victims: The impact of welfare states (the case of the Netherlands) on refugees," *Journal of Refugee Studies*, vol. 18, no. 2, pp. 181–198.

Ghosh, P 2010, *Bhutanese refugees: A forgotten saga*, Minerva, Kolkata.

Ghosn, F, Braithwaite, A and Chu, TS 2019, "Violence, displacement, contact, and attitudes toward hosting refugees," *Journal of Peace Research*, vol. 56, no. 1, pp.118–133.

Giddens, A 1985, *The nation-state and violence*, University of California Press, Berkeley.

Gilroy, P 1990, "Nationalism, history and ethnic absolutism," *History Workshop*, vol. 30, no. 1, pp. 114–120.

Goffman, E 1959, *The presentation of self in everyday life*, Doubleday, Garden City.

Goode, J and Maskovsky, J (eds.) 2001, *New poverty studies: The ethnography of power, politics and impoverished people in the United States*, New York University Press, New York.

Gordon, M 2012, "Refugees are 'boat people' to most, UN survey finds," *The Canberra Times*, viewed 11 November 2015, http://www.canberratimes.com.au/federal-politics/political-news/refugees-are-boat-people-to-most-un-survey-finds-20120617-20ide.html.

Greig, A, Lewins, F and White, K 2003, *Inequality in Australia*, Cambridge University Press, Massachusetts.

Grindal, BT and Salamone, FA (eds.) 2006, *Bridges to humanity: Narratives on fieldwork and friendship*, Waveland Press Inc., Long Grove.

Gupta, A and Ferguson, J 1992, "'Beyond culture': Space, identity, and the politics of difference," *Cultural Anthropology*, vol. 7, no. 1, pp. 6–23.

Gupta, R 1975, "Sikkim: The merger with India," *Asian Survey*, vol. 15, no. 9, pp. 786–798.

Hage, G 1996, "Nationalist anxiety or the fear of losing your other," *The Australian Journal of Anthropology*, vol. 7, no. 2, pp. 121–140.

Hage, G 1998, *White nation: Fantasies of white supremacy in a multicultural society*, Pluto Press, Annandale.

Hage, G 2002, "Multiculturalism and white paranoia in Australia," *Journal of International Migration and Integration/Revue de l'integration et de la migration internationale*, vol. 3, no. 3–4, pp. 417–437.

Hage, G 2003, *Against paranoid nationalism: Searching for hope in a shrinking society*, Pluto Press, Annandale.

Hagerty, DT 1991, "India's regional security doctrine," *Asian Survey*, vol. 31, no. 4, pp. 351–363.

Hall, S 2006, "The west and the rest: Discourse and power." In CA Roger and C Andersen (eds.), *The Indigenous experience: Global perspectives*, Canadian Scholar's Press Inc., Toronto, pp. 165–173.

Halper, T 1973, "The poor as pawns: The new 'deserving poor' & the old," *Polity*, vol. 6, no. 1, pp. 71–86.

Harrell-Bond, B 1986, *Imposing aid: Emergency assistance to refugees*, Oxford University Press, Oxford.

Harrell-Bond, B 1999, "The experience of refugees as recipients of aid." In A Ager (ed.), *Refugees: Perspectives on the experience of forced migration*, Pinter, New York, pp. 136–168.

Harrell-Bond, B 2002, "Can humanitarian work with refugees be humane?" *Human Rights Quarterly*, vol. 24, no. 1, pp. 51–85.

Harrell-Bond, B and Voutira, E 2007, "In search of 'invisible' actors: Barriers to access in refugee research," *Journal of Refugee Studies*, vol. 20, no. 2, pp. 281–298.

Harris, M 1966, "The cultural ecology of India's sacred cattle [and comments and replies]," *Current Anthropology*, vol. 7, no. 1, pp. 51–66.

Healy, P and Bosman, J 2015, "G.O.P. governors vow to close doors to Syrian refugees," *The New York Times*, 16 November, viewed 23 November 2015, www.nytimes.com/2015/11/17/us/politics/gop-governors-vow-to-close-doors-to-syrian-refugees.html.

Heyman, JM and Symons J 2015, "Borders." In D Fassin (ed.), *A companion to moral anthropology*, John Wiley and Sons, Inc., Oxford, pp. 540–557.

Höijer, B 2004, "The discourse of global compassion: The audience and media reporting of human suffering," *Media, Culture & Society*, vol. 26, no. 4, pp. 513–531.

Horst, C 2006, *Transnational nomads: How Somalis cope with refugee life in the Dadaab camps of Kenya*, Berghahn Books, New York.

Hugo, G 1986, *Australia's changing population: Trends and implications*, Oxford University Press, Melbourne.

Hugo, G 1992, "Knocking at the door: Asian immigration to Australia," *Asian and Pacific Migration Journal*, vol. 1, no. 1, pp. 100–144.

Human Rights Committee 2000, *General comment 28, equality of rights between men and women (article 3)*, CCPR/C/21/Rev.1/Add.10, United Nations Document, Geneva.

Human Rights Watch 2003, *Bhutan/Nepal: Trapped by inequality: Bhutanese refugee women in Nepal*, Report, Human Rights Watch, New York.

Hutchinson, M and Dorsett, P 2012, "What does the literature say about resilience in refugee people? Implications for practice," *Journal of Social Inclusion*, vol. 3, no. 2, pp. 55–78.

Hutt, M 1994, *Bhutan: Perspectives on conflict and dissent*, Kiscadale Publications, Gartmore.

Hutt, M 1996, "Ethnic nationalism, refugees and Bhutan," *Journal of Refugee Studies*, vol. 9, no. 4, pp. 397–420.

Hutt, M 1997, "Being Nepali without Nepal: Reflections on a South Asian diaspora." In DN Gellner, J Pfaff-Czarnecka and J Whelpton (eds.), *Nationalism and ethnicity in a Hindu kingdom: The politics of culture in contemporary Nepal*, Overseas Publishers Association, Harwood, pp. 101–144.

Hutt, M 2003, *Unbecoming citizens: Culture, nationhood, and the flight of refugees from Bhutan*, Oxford University Press, New Delhi.

Hutt, M 2005, "The Bhutanese refugees: Between verification, repatriation and royal realpolitik," *Peace and Democracy in South Asia*, vol. 1, no. 1, pp. 44–56.

Hyndman, J 2000, *Managing displacement: Refugees and the politics of humanitarianism*, University of Minnesota Press, Minneapolis.

Hyndman, J 2010, "Introduction: The feminist politics of refugee migration," *Gender, Place and Culture*, vol. 17, no. 4, pp. 453–459.

Hyndman, J and Giles, W 2011, "Waiting for what? The feminization of asylum in protracted situations," *Gender, Place & Culture*, vol. 18, no. 3, pp. 361–379.

Inglis, C 2009, "Multicultural education in Australia: Two generations of evolution." In JA Banks (ed.), *The Routledge international companion to multicultural education*, Routledge, New York, pp. 109–120.

Inglis, C and Wu, CT 1992, "The 'new' migration of Asian skills and capital to Australia." In C Inglis and CT Wu (eds.), *Asians in Australia: The dynamics of migration and settlement*, Monograph no. 130, Institute of South Asia Studies, Singapore, pp. 193–230.

International Organization for Migration (IOM) 2015a, "Resettlement of refugees from Bhutan tops 100,000," IOM, viewed 24 May 2023, https://www.iom.int/news/resettlement-refugees-bhutan-tops-100000.

International Organization for Migration (IOM) 2015b, *Europe/Mediterranean migration response*, IOM Preparedness and Response Division, Geneva.

Ismail, Y 2006, "Fingerprints mark new direction in refugee registration," *News Stories*, 30 November, United Nations High Commissioner for Refugees, Geneva.

Janmyr, M 2019, "The 1951 refugee convention and non-signatory states: Charting a research agenda," *International Journal of Refugee Law*, vol. 33, no. 2, pp. 188–213.

Johnson, HL 2011, "Click to donate: Visual images, constructing victims and imagining the female refugee," *Third World Quarterly*, vol. 32, no. 6, pp. 1015–1037.

Johnson, HL 2014, *Borders, asylum and global non-citizenship: The other side of the fence*, Cambridge University Press, Cambridge.

Jupp, J 1984, *Ethnic politics in Australia*, G. Allen & Unwin, Sydney.
Jupp, J 1995, "From 'white Australia' to 'part of Asia': Recent shifts in Australian immigration policy towards the region," *International Migration Review*, vol. 29, no. 1, pp. 207-228.
Jupp, J 2007, *From white Australia to Woomera: The story of Australian immigration*, 2nd edition, Cambridge University Press, Port Melbourne.
Kapoor, I 2013, *Celebrity humanitarianism: The ideology of global charity*, Routledge, London.
Karan, PP and Jenkins, WM 1963, *The Himalayan kingdoms: Bhutan, Sikkim, and Nepal*, D. Van Nostrand Company, New Jersey.
Kenny, P and Lockwood-Kenny, K 2011, "A mixed blessing: Karen resettlement in the United States," *Journal of Refugee Studies*, vol. 24, no. 2, pp. 217-238.
King, R 1999, "Orientalism and the modern myth of 'Hinduism,'" NUMEN *International Review for the History of Religions*, vol. 46, no. 2, pp. 146-185.
Kingston, B 1975, *My wife, my daughter, and poor Mary Ann: Women and work in Australia*, Thomas Nelson, Melbourne.
Lai, D 2000, *The poverty of "development economics,"* MIT Press, Cambridge.
Lareau, A 1996, "Common problems in field work: A personal essay." In A Lareau, J Shultz and JJ Shultz (eds.), *Journeys through ethnography: Realistic accounts of fieldwork*, Westview Press, Boulder, pp. 196-236.
Lee, TL 1998, "Refugees from Bhutan: Nationality, statelessness and the right to return," *International Journal of Refugee Law*, vol. 10, no. 1/2, pp. 118-155.
Leech, BL 2002, "Asking questions: Techniques for semi-structured interviews," *Political Science & Politics*, vol. 35, no. 4, pp. 665-668.
Leonard, K 2009, "Changing places: The advantages of multi-sited ethnography." In MA Falzon (ed.), *Multi-sited ethnography: Theory, praxis, and locality in contemporary research*, Ashgate Publishing Limited, Surrey, pp. 165-179.
Levi, W 1959, "Bhutan and Sikkim: Two buffer states," *The World Today*, vol. 15, no. 12, pp. 492-500.
Lewis, B 2002, *What went wrong? Western impact and Middle Eastern response*, Oxford University Press, New York.
Lewis, H 2010, "Community moments: Integration and transnationalism at refugee parties and events," *Journal of Refugee Studies*, vol. 23, no. 4, pp. 571-588.
Limbu, B 2009, "Illegible humanity: The refugee, human rights, and the question of representation," *Journal of Refugee Studies*, vol. 22, no. 3, pp. 257-282.
Loescher, G, Betts, A and Milner, J 2008, *The United Nations High Commissioner for Refugees (UNHCR): The politics and practice of refugee protection into the 21st century*, Routledge, New York.

Lokot, M 2019, "The space between us: Feminist values and humanitarian power dynamics in research with refugees," *Gender & Development*, vol. 27, no. 3, pp. 467–484.

Long, L 1993, *Ban Vinai, the refugee camp*, Columbia University Press, New York.

Lubet, S 2018, *Interrogating ethnography: Why evidence matters*, Oxford University Press, New York.

Mackenzie, C, McDowell, C and Pittaway, E 2007, "Beyond 'do no harm': The challenge of constructing ethical relationships in refugee research," *Journal of Refugee Studies*, vol. 20, no. 2, pp. 299–319.

Malagodi, M 2021, "Holy cows and constitutional nationalism in Nepal," *Asian Ethnology*, vol. 80, no. 1, pp. 93–120.

Malkki, L 1995a, *Purity and exile: Violence, memory, and national cosmology among Hutu refugees in Tanzania*, University of Chicago Press, Chicago.

Malkki, L 1995b, "Refugees and exile: From 'refugee studies' to the national order of things," *Annual Review of Anthropology*, vol. 24, pp. 495–523.

Malkki, L 1996, "Speechless emissaries: Refugees, humanitarianism, and dehistoricization," *Cultural Anthropology*, vol. 11, no. 3, pp. 377–404.

Manne, R 1987, *The Petrov affair: Politics and espionage*, Pergamon Press, Sydney.

Marcus, G 1995, "Ethnography in/of the world system: The emergence of multi-sited ethnography," *Annual Review of Anthropology*, vol. 24, pp. 95–117.

Marlowe, JM 2010, "Beyond the discourse of trauma: Shifting the focus on Sudanese refugees," *The Journal of Refugee Studies*, vol. 23, no. 2, pp.183–198.

McAdam, J 2013, "Australia and asylum seekers," *The Journal of Refugee Studies*, vol. 25, no. 3, pp. 435–448.

McAdam, J and Chong, F 2019, *Refugee rights and policy wrongs: A frank, up-to-date guide by experts*, University of New South Wales Press, Sydney.

McCarthy, P 2005, "After 50 years is Elizabeth a success?" *Stateline*, radio transcript, ABC Radio National, 18 November, viewed 17 November 2015, http://www.abc.net.au/stateline/sa/content/2005/s1512761.htm.

Mckay, FH, Thomas, SL and Kneebone, S 2012, "'It would be okay if they came through the proper channels': Community perceptions and attitudes toward asylum seekers in Australia," *Journal of Refugee Studies*, vol. 25, no. 1, pp. 113–133.

McKinnon, SL 2008, "Unsettling resettlement: Problematizing 'Lost boys of Sudan' resettlement and identity," *Western Journal of Communication*, vol. 72, no. 4, pp. 397–414.

McMaster, D 2002, "Asylum-seekers and the insecurity of a nation," *Australian Journal of International Affairs*, vol. 56, no. 2, pp. 279–290.

Mence, V, Gangell, S and Tebb, R 2015, *A history of the Department of Immigration: Managing migration to Australia*, Department of Immigration and Border Protection, The Australian Government, Canberra.

Michaels, A 2004, *Hinduism: Past and present*, Princeton University Press, Princeton.

Milanovic, B 2011, *Worlds apart: Measuring international and global inequality*, Princeton University Press, Princeton.

Morrison, S 2013a, "Conduct of behaviour asylum seekers," 20 December, Operation Sovereign Borders Press Conference, Canberra.

Morrison, S 2013b, "Paramasala festival: Citizenship ceremony," 7 October, Paramatta Riverside Theatre, Paramatta.

Morsink, J 1999, *The Universal Declaration of Human Rights: Origins, drafting, and intent*, University of Pennsylvania Press, Philadelphia.

Muehlmann, S 2009, "How do real Indians fish? Neoliberal multiculturalism and contested Indigeneities in the Colorado delta," *American Anthropologist*, vol. 111, no. 4, pp. 468–479.

Muggah, R 2005, "Distinguishing means and ends: The counterintuitive effects of UNHCR's community development approach in Nepal," *Journal of Refugee Studies*, vol. 18, no. 2, pp. 151–164.

Müller, TR 2013, "'The Ethiopian famine' revisited: Band Aid and the anti-politics of celebrity humanitarian action," *Disasters*, vol. 37, no. 1, pp. 61–79.

Mullick, SB 2001, "Indigenous peoples and electoral politics in India: An experience of incompatibility." In K Wessendorf (ed.), *Challenging politics: Indigenous peoples' experiences with political parties and elections*, International Work for Indigenous Affairs, Copenhagen, pp. 94–145.

Munro, A 1999, "Humanitarianism and conflict in a post-Cold War world," *Revue Internationale de la Croix-Rouge/International Review of the Red Cross*, vol. 81, no. 835, pp. 463–476.

Murray Li, T 2000, "Articulating Indigenous identity in Indonesia: Resource politics and the tribal slot," *Comparative Studies in Society and History*, vol. 42, no. 1, pp. 149–179.

Nah, AM 2016, "Networks and norm entrepreneurship amongst local civil society actors: Advancing refugee protection in the Asia Pacific region," *The International Journal of Human Rights*, vol. 20, no. 2, pp. 223–240.

Nath, L 2005, "Migrants in flight: Conflict-induced internal displacement of Nepalis in northeast India," *Peace and Democracy in South Asia*, vol. 1, no. 1, pp. 56–72.

Neikirk, AM 2017, "Expectations of vulnerability in Australia," *Forced Migration Review*, vol. 54, pp. 63–65.

Neikirk, AM 2018, "A moral marriage: Humanitarian values and the Bhutanese refugees," *Journal of Refugee Studies*, vol. 31, no. 1, pp. 63–81.

Nelson, A and Stam, K 2021, "Bhutanese or Nepali? The politics of ethnonym ambiguity," *South Asia: Journal of South Asian Studies*, 1–18.

Neumann, K 2015, *Across the seas: Australia's response to refugees: A history*, Black Inc., Melbourne.

Neumann, K and Tavan, G 2009, *Does history matter? Making and debating citizenship, immigration and refugee policy in Australia and New Zealand*, ANU Press, Canberra.

Nikolic-Ristanovic, V 2003, "Refugee women in Serbia—Invisible victims of war in the former Yugoslavia," *Feminist Review*, vol. 73, no. 1, pp. 104–113.

Nolan, D, Farquharson, K, Politoff, V and Marjoribanks, T 2011, "Mediated multiculturalism: Newspaper representations of Sudanese migrants in Australia," *Journal of Intercultural Studies*, vol. 32, no. 6, pp. 655–671.

O'Brien, M 2010, "Fears of 'demographic inundation' behind Bhutan's refugee crisis," *East Bay Times*, 28 August, viewed 31 March 2022, http://www.eastbaytimes.com/san-leandro/ci_15906108?source=pkg.

Ogura, K 2007, "Maoists, people, and the state as seen from Rolpa and Rukum." In H Ishii, D Gellner and K Nawa (eds.), *Political and social transformations in North India and Nepal: Social dynamics in Northern South Asia*, Manohar Publication, New Delhi, pp. 435–475.

Oldenburg, VT 2002, *Dowry murder: The imperial origins of a cultural crime*, Oxford University Press, Oxford.

Oliver, A 2014, *The Lowy Institute poll 2014*, Lowy Institute for International Policy, Sydney.

Ong, A 1987, *Spirits of resistance and capitalist discipline: Factory women in Malaysia*, State University of New York Press, Albany.

Ong, A 1999, *Flexible citizenship: The cultural logics of transnationality*, Duke University Press, Durham.

Ong, A 2003, *Buddha is hiding: Refugees, citizenship, and the New America*, University of California Press, Berkeley.

Orford, A 2010, "The passions of protection: Sovereign authority and humanitarian war." In D Fassin and M Pandolfi (eds.), *Contemporary states of emergency: The politics of military and humanitarian interventions*, Zone Books, New York, pp. 335–356.

Ortner, SB 2006, *Anthropology and social theory: Culture, power, and the acting subject*, Duke University Press, Durham.

Owens, P 2009, "Reclaiming 'bare life'? Against Agamben on refugees," *International Relations*, vol. 23, no. 4, pp. 567–582.

Palfreeman, AC 1967, *The administration of the White Australia Policy*, Melbourne University Press, Melbourne.

Peel, M 2003, *The lowest rung: Voices of Australian poverty*, Cambridge University Press, Cambridge.

Pettman, JJ 1995, "Race, ethnicity, and gender in Australia." In D Stasiulis and N Yuval-Davis (eds.), *Unsettling settler societies: Articulations of gender, race, ethnicity and class*, vol. 11, Sage, London, pp. 65–93.

Pfaff-Czarnecka, J 1997, "Vestiges and visions: Cultural change in the process of nation-building in Nepal." In DN Gellner, J Pfaff-Czarnecka and J Whelpton (eds.), *Nationalism and ethnicity in a Hindu kingdom: The politics of culture in contemporary Nepal*, Overseas Publishers Association, Harwood, pp. 419–470.

Phillips, J, Klapdor, M and Simon-Davies, J 2010, *Migration to Australia since federation: A guide to the statistics*, Department of Parliamentary Services, The Australian Government, Canberra.

Phuntsho, K 2013, *The history of Bhutan*, Random House India, New Delhi.

Pickering, S 2001, "Common sense and original deviancy: News discourses and asylum seekers in Australia," *Journal of Refugee Studies*, vol. 14, no. 2, pp. 169–186.

Piven, FF and Cloward, RA 1971, *Regulating the poor: The functions of social welfare*, Vintage, New York.

Povinelli, EA 2002, *The cunning of recognition: Indigenous alterities and the making of Australian multiculturalism*, Duke University Press, Durham.

Prakash, V 2008, *Terrorism in India's north-east: A gathering storm*, vol. 2, Kalpaz Publications, New Delhi.

Pupavac, V 2008, "Refugee advocacy, traumatic representations and political disenchantment," *Government and Opposition*, vol. 43, no. 2, pp. 270–292.

Rajaram, PK 2002, "Humanitarianism and representations of the refugee," *Journal of Refugee Studies*, vol. 15, no. 3, pp. 247–264.

Redfield, P 2013, *Life in crisis: The ethical journey of Doctors Without Borders*, University of California Press, Berkeley.

Reilly, R 1994, "Life and work in the refugee camps of southeast Nepal." In M Hutt (ed.), *Bhutan: Perspectives on conflict and dissent*, Kiscadale Publications, Gartmore, 129–140.

Rennie, DF 1866, *Bhutan and the story of the Duar War*, J. Murray, London.

Ridderbos, K 2007, *Last hope: The need for durable solutions for Bhutanese refugees in Nepal and India*, vol. 9, no. 7, Human Rights Watch, New York.

Rist, G 2014, *The history of development: From western origins to global faith*, Zed Books, London.

Robbins, P 1999, "Meat matters: Cultural politics along the commodity chain in India," *Cultural Geographies*, vol. 6, no. 4, pp. 399–423.

Rose, LE 1963, "The Himalayan border states: 'Buffers' in transition," *Asian Survey*, vol. 3, no. 2, pp. 116–122.

Rose, N 1996, "The death of the social? Re-figuring the territory of government," *International Journal of Human Resource Management*, vol. 25, no. 3, pp. 327–356.

Rose, N 1999, *Powers of freedom: Reframing political thought*, Cambridge University Press, Cambridge.

Rose, N 2000, "Government and control," *British Journal of Criminology*, vol. 40, no. 2, pp. 321–339.

Rose, N 2006, "Governing 'advanced' liberal democracies." In A Sharma and A Gupta (eds.), *The anthropology of the state: A reader*, John Wiley & Sons, Somerset, pp. 144–162.

Rose, N 2017, "Still 'like birds on the wire'? Freedom after neoliberalism," *Economy and Society*, vol. 46, no. 3–4, pp. 303–323.

Royal Government of Bhutan 1953, *First general assembly meeting: Official minutes*, The General Assembly, Thimphu.

Royal Government of Bhutan 1992, *Anti-national activities in Southern Bhutan: An update on the terrorist movement*, Department of Information, Thimphu.

Royal Government of Bhutan 1993, *The Southern Bhutan problem: Threat to a nation's survival*, Ministry of Home Affairs, Thimphu.

Salter, MB 2008, "When the exception becomes the rule: Borders, sovereignty, and citizenship," *Citizenship Studies*, vol. 12, no. 4, pp. 365–380.

Sarkar, R and Ray, I 2007, "Political scenario in Bhutan during 1774–1906: An impact analysis on trade and commerce," *Journal of Bhutan Studies*, vol. 17, pp. 1–21.

Saul, B 2000, "Cultural nationalism, self-determination and human rights in Bhutan," *International Journal of Refugee Law*, vol. 12, no. 3, pp. 321–353.

Schmitt, C 1985, *Political theology: Four chapters on the concept of sovereignty*, University of Chicago Press, Chicago.

Schuster, L 2003, "Common sense or racism? The treatment of asylum-seekers in Europe," *Patterns of Prejudice*, vol. 37, no. 3, pp. 233–256.

Selznick, P 1994, *The moral commonwealth: Social theory and the promise of community*, University of California Press, Berkeley.

Shneiderman, S 2010, "Creating 'civilized' communists." In D Gellner (ed.), *Varieties of activist experience: Civil society in South Asia*, Sage, New Delhi, pp. 46–80.

Shrestha, N 1995, "Becoming a development category." In J Crush (ed.), *Power of development*, Routledge, London, pp. 259–270.

Skran, CM 1992, "The international refugee regime: The historical and contemporary context of international responses to asylum problems," *Journal of Policy History*, vol. 4, no. 1, pp. 8–35.

Smith, M 2004, "Warehousing refugees," *World Refugee Survey*, vol. 38, no. 1, pp. 38–56.

Spradley, JP 1979, *The ethnographic interview*, Holt, Rinehart and Winston, New York.

Squire, V 2009, *The exclusionary politics of asylum*, Palgrave Macmillan Ltd., New York.

Stamatov, P 2013, *The origins of global humanitarianism: Religion, empires, and advocacy*, Cambridge University Press, Cambridge.

Stein, BN 1981, "The refugee experience: Defining the parameters of a field of study," *International Migration Review*, pp. 320-330.

Stewart, F, Ranis, G and Samman, E 2018, *Advancing human development: Theory and practice*, Oxford University Press, Oxford.

Stirr, A 2010, "'May I elope': Song words, social status, and honor among female Nepali Dohori singers," *Ethnomusicology*, vol. 54, no. 2, pp. 257-280.

Stone, L 1978, "Cultural repercussions of childlessness and low fertility in Nepal," *Contributions to Nepalese Studies*, vol. 5, no. 2, pp. 7-36.

Stubbs, P 1996, "Nationalisms, globalization and civil society in Croatia and Slovenia," *Research in Social Movements, Conflicts and Change*, vol. 19, no. 1, pp. 1-26.

Suhrke, A and Zolberg, AR 1989, "Beyond the refugee crisis: Disengagement and durable solutions for the developing world," *Migration*, vol. 5, pp. 69-119.

Szczepanikova, A 2013, "Beyond 'helping': Gender and relations of power in non-governmental assistance to refugees," *Journal of International Women's Studies*, vol. 11, no. 3, pp. 19-33.

Taylor, C 2011, *Dilemmas and connections: Selected essays*, Harvard University Press, Cambridge.

Thinley, JY 1994, "A kingdom besieged." In M Hutt (ed.), *Bhutan: Perspectives on conflict and dissent*, Kiscadale Publications, Gartmore, pp. 43-76.

Thomas, DA and Kamari Clarke, M 2013, "Globalization and race: Structures of inequality, new sovereignties, and citizenship in a neoliberal era," *Annual Review of Anthropology*, vol. 42, pp. 305-325.

Ticktin, M 2006, "Where ethics and politics meet: The violence of humanitarianism in France," *American Ethnologist*, vol. 33, no. 1, 33-49.

Ticktin, M 2011, *Casualties of care: Immigration and the politics of humanitarianism in France*, University of California Press, Berkeley.

Tucker, V 1999, "The myth of development: A critique of a Eurocentric discourse." In R Munck and D O'Hearn (eds.), *Critical development theory: Contributions to a new paradigm*, Zed Books, London, pp. 1-26.

United Nations Development Programme (UNDP) 1987, *UNHCR/UNDP co-operation with regard to development activities affecting refugees and returnees*, Office of the United Nations High Commissioner for Refugees, Geneva.

United Nations High Commissioner for Refugees (UNHCR) 1992, *Report of the United Nations High Commissioner for Refugees*, Office of the United Nations High Commissioner for Refugees, Geneva.

United Nations High Commissioner for Refugees (UNHCR) 1997, *Memorandum of understanding between UNDP and UNHCR in Rwanda*, Office of the United Nations High Commissioner for Refugees, Geneva.

United Nations High Commissioner for Refugees (UNHCR) 2005, "Module 8: Vulnerable groups," *Reach out refugee protection training project*, Office of the United Nations High Commissioner for Refugees, Geneva.

United Nations High Commissioner for Refugees (UNHCR) 2008, *A community-based approach in UNHCR operations*, Office of the United Nations High Commissioner for Refugees, Geneva.

United Nations High Commissioner for Refugees (UNHCR) 2009, "Nepal," *UNHCR global appeal 2010–11*, Office of the United Nations High Commissioner for Refugees, Geneva.

United Nations High Commissioner for Refugees (UNHCR) 2011, *Resettlement handbook*, Office of the United Nations High Commissioner for Refugees, Geneva.

United Nations High Commissioner for Refugees (UNHCR) 2015, *2015 contributions to UNHCR programmes*, Office of the United Nations High Commissioner for Refugees, Geneva.

United Nations High Commissioner for Refugees (UNHCR) 2019, *Refugee education 2030: A strategy for refugee inclusion*, Office of the United Nations High Commissioner for Refugees, Geneva.

United Nations High Commissioner for Refugees (UNHCR) 2022, *Global trends: Forced displacement in 2020*, Office of the United Nations High Commissioner for Refugees, Geneva.

United Nations High Commissioner for Refugees (UNHCR), United Nations Office for the Coordination of Humanitarian Action, Survey Department Nepal 2015, *Nepal: Administrative map*, Office of the United Nations High Commissioner for Refugees, Geneva.

US Committee for Refugees and Immigrants 1988, *World refugee survey: 1987 in review*, US Committee for Refugees and Immigrants, Arlington.

US Committee for Refugees and Immigrants 2002, *US committee for refugees world refugee survey 2002—Bhutan*, US Committee for Refugees and Immigrants, Arlington.

US Department of State 1990, *Country reports—Bhutan*, US Department of State, Washington, DC.

US Department of State 2008, *Bhutan: International religious freedom report*, US Department of State, Washington, DC.

Valtonen, K 2004, "From the margin to the mainstream: Conceptualizing refugee settlement processes," *Journal of Refugee Studies*, vol. 17, no. 1, pp. 70–96.

Van Driem, G 1994, "Language policy in Bhutan." In M Aris and M Hutt (eds.), *Bhutan: Aspects of culture and development*, Kiscadale Publications, Gartmore, pp. 87–105.

Watson, SD 2011, "Impugning the humanitarian defence," *International Migration*, vol. 52, no. 2, pp. 353–373.

Welch, M and Schuster, L 2005, "Detention of asylum seekers in the UK and USA deciphering noisy and quiet constructions," *Punishment & Society*, vol. 7, no. 4, pp. 397–417.

Westoby, P 2008, "Developing a community-development approach through engaging resettling Southern Sudanese refugees within Australia," *Community Development Journal*, vol. 43, no. 4, pp. 483–495.

Whelpton, J 1997, "Political identity in Nepal: State, nation, and community." In DN Gellner, J Pfaff-Czarnecka and J Whelpton (eds.), *Nationalism and ethnicity in a Hindu kingdom: The politics of culture in contemporary Nepal*, Overseas Publishers Association, Harwood, pp. 39–78.

Whelpton, J 2005, *A history of Nepal*, Cambridge University Press, Cambridge.

Whelpton, J, Gellner, DN and Pfaff-Czarnecka, J 2008, "New Nepal, new ethnicities: Changes since the mid 1990s," In DN Gellner, J Pfaff-Czarnecka and J Whelpton (eds.), *Nationalism and ethnicity in Nepal*, Vajra Books, Kathmandu, pp. xvii–xlviii.

White, JC 1909, *Sikhim & Bhutan: Twenty-one years on the north-east frontier, 1887–1908*, Longmans, Green, New York.

Whitecross, RW 2010, "Intimacy, loyalty and state formation." In S Thiranagama and T Kelly (eds.), *Traitors: Suspicion, intimacy, and the ethics of state-building*, University of Pennsylvania Press, Philadelphia, pp. 68–88.

Whitlam, G 1985, *The Whitlam government: 1972–1975*, Viking Penguin Books Australia Ltd., Ringwood.

Wolf, ER and Eriksen, TH 2010, *Europe and the people without history*, 2nd edition, University of California Press, Berkeley.

Woodley, B and Newton, D 1987, "I am Australian," Warner/Chappell Music Australia, North Ryde.

World Food Programme 2006, *Report of UNHCR/WFP joint assessment mission 29 May 2006–9 June 2006*, WFP Regional Bureau for Asia, Bangkok.

Yadav, P 2016, *Social transformation in post-conflict Nepal: A gender perspective*, Routledge, New York.

Yarwood, AT 1958, "The dictation test—historical survey," *The Australian Quarterly*, vol. 30, no. 2, pp. 19–29.

Zakus, D, Skinner, J and Edwards, A 2009, "Social capital in Australian sport," *Sport in Society*, vol. 12, no. 7, pp. 986–998.

Zetter, R 1991, "Labelling refugees: Forming and transforming a bureaucratic identity," *Journal of Refugee Studies*, vol. 4, no. 1, pp. 39–62.

Zetter, R 2007, "More labels, fewer refugees: Remaking the refugee label in an era of globalization," *Journal of Refugee Studies*, vol. 20, no. 2, pp. 172–192.

Zolberg, AR, Suhrke, A and Aguayo, S 1989, *Escape from violence: Conflict and the refugee crisis in the developing world*, Oxford University Press, New York.

Legislation

Bhutan Citizenship Act 1985 (Bhutan).
Citizenship Act 1955 Republic of India (India).
Commonwealth Racial Discrimination Act of 1975 (Australia).
Convention on the Rights of the Child 1989 (UNICEF).
Convention Relating to the Status of Refugees 1951 (UN General Assembly).
Declaration on the Rights of Indigenous Peoples 2007 (UN General Assembly).
Family Law Act 1975 (Australia).
Foreigners Act 1946 (India).
Immigration Restriction Act 1901 (Australia).
International Convention on the Elimination of All Forms of Racial Discrimination 1965 (UN General Assembly).
Marriage Act of Bhutan 1980 (Bhutan).
Migration Act 1958 (Australia).
Muluki Ain 1854 (1990 Amendment) (Nepal).
National Foundation for Development of Indigenous Nationalities Act 2002 (Nepal).
Nationality Law of Bhutan 1958 (Bhutan).
Nepal Citizenship Act 1964 [1967] (Nepal).
Protocol against the Smuggling of Migrants by Land, Sea and Air 2000 (UN General Assembly).
Protocol Relating to the Status of Refugees 1967 (UN General Assembly).
Treaty of Friendship 1949 (India).
Treaty of Peace and Friendship between the Government of India and the Government of Nepal 1950 (Nepal).
United Nations Convention against Transnational Organized Crime 2000 (UN General Assembly).
Universal Declaration of Human Rights 1948 (UN General Assembly).
War-Time Refugees Removal Act 1949 (Australia).

Index

Page numbers in italics represent maps; t represents tables.

Abbott, Tony, 121, 205
Aboriginal people, 115
absolution, 211–13
activism. *See* political activism
Adamson, FB., 33–34
Adelaide, Australia, 155, 159, 165
Africa, 5, 6
aid dependency, 9
Amnesty International, 42
Arendt, H., 210
Asia, 6
astrology, 76
asylum seekers, 113–15, 120–23, 205–06, 214. *See also* refugees
Australia: charity donations, 20; as diverse, 126; Harmony Day, 135–36; immigrant history, 115–18; minorities as contained, 21–22; multiculturalism, 124–27, 135–36; refugee history, 118–22; refugees thanking, 27; refugees vs. migrants, 20–21; social welfare, 127, 189–90; working

class, 129. *See also* Salisbury, Australia
Australian Citizenship: Our Common Bond (guide), 128
Australia Refugee Association (ARA), 130
Australia resettlement: overview, 19–20; alternative relationships, 188; asylum seekers vs. referred refugees, 113–15, 205–06; BAASA school, 156–58; and Bhutanese identity, 148–49; citizenship, 128, 199–203; clubs, 179–81; and community as egalitarian, 167; cultural problems, 134; deference of capability, 157–58; and dowry, 146, 147–48; and employment, 149–53; and English language, 156, 157–58; as ethical act, 123; family structure of Bhutanese, 143–44; ghettoization of Bhutanese, 133; and governance, 146, 149; government as family, 144; and government social standing, 133–34, 136, 140; initiatives by Bhutanese,

239

198–99; integration overview, 127, 130–31; managing refugees, 134, 136, 139, 154–58, 212; narratives of Bhutanese, 191–95, 197; neighborhoods, 165–66, 186; performance of culture, 138–39; and polygamy, 146–47, 148; priorities, 189; resources, 127, 130–31; returning to camps, 201–02; scarf-giving, 174–75; settlement expectations, 108; special treatment perception, 199–200; Thursday meetings, 136–37, 139, 162, 174, 177; "tradition" used to define, 198; and trauma, 152–53, 194–95; trauma and worthiness, 21, 122–23, 189; women focus, 141–50; yoga classes, 137. *See also* caste system in Australia; dowry; integration; polygamy

Australia Values Statement (publication), 128

autonomy, 157–58

Banki, S., 151
barbed wire fences, 53–54
Barnett, M., 10
Baumann, G., 170
beef, 103–04
Belgandi camps. *See* camps
Bhutan: borders, 39; government attitude to refugees, 56; history of, 33, 35–42, 43, 192; and India, 43–45; as isolated, 34; and Nepal, 46
Bhutanese (overview): becoming, 55–57; and caste system in research, 26; culture of, 33, 39, 55–56, 134, 138–39; as elite refugees, 2, 109–10; elites, 37–38 (*see also* Drukpa people); as esteemed, 34; family structure, 143–44; identity and nation building, 39–41; identity in Australia, 148–49; as molded, 22; as pawns, 42, 45; repatriation, 47; as righteous, 60–61; younger generation, 88. *See also* dowry; marriage; polygamy

Bhutanese Australia Association of South Australia (BAASA), 155–61, 181, 198

birth control, 67
Bodoland Movement, 44
book overview, 30–31, 212–13
borders in Bhutan, 39
Bornstein, E., 212
Brahmins (overview), 91–92, 94–95, 165, 168–70. *See also* caste system
Britain, 34–38
Brohman, J., 10
Buddhism, 34, 36, 39
Buddism, 105
Burkert, C., 92
Burundi, 38

camps: overview, 47–50; and Australian resettlement, 19; becoming Bhutanese, 55–57; and caste system, 61–64, 66, 93–97, 102, 170–71; cleaning, 66; and community projects, 158; and contraceptives, 67; conversion to Christianity, 100–01; in India, 44–45, 48; and Indigenous identity, 105–07; and information exchange, 25; internalizing values, 67, 69, 70–71, 75, 77, 98; leaders in, 25–26, 57, 63–64, 70–71, 83–84; map of, 48; and marriage, 76;

and mixed marriages, 99–101; and morality, 11; and mythic histories, 17–18; National Day, 55, 81; in Nepal (overview), 47–48, 48; Nepalese workers, 58, 69–70, 85; as normalized, 211–12; number of people in, 213; paid work of refugees, 94; refugees vs. locals, 57–60, 61, 85; and refugee warriors, 7; reinvention in, 68–69; resources, 57–59, 61, 85; return to, 201–02; Sanischare camp, 49; scarf-giving, 174; schools in, 61–62, 89; security in, 53–54; transformation in (overview), 54–55, 209–10; transportation to Damak, 86–87; and trauma, 153; volunteerism, 69, 70–71, 85, 86; working refugees, 58, 69–70, 85, 86–89, 90, 94. *See also* transformation

Caritas-Nepal, 61–62, 89, 99

caste system: overview, 26, 64n1, 91–92, 102–03, 178n1; assumptions, 91–92; attention to, 96–97; and BAASA, 156, 159–60; blaming Nepalese, 65; in camps, 61–64, 66, 93–97, 102, 170–71; and community, 98–102; contamination, 96–97; defined, 92; and democracy, 196–97; and education, 61–63, 64–65, 98, 99; and English language, 180; and ethnicity, 91; hiding, 96–97, 102, 195–96; and Hinduism, 196; and human rights, 61; and Indigenous identity, 107; initial Bhutan refugees, 46; and marriage, 76, 98–104, 177, 183; Nepal social hierarchy table, 93t; and regulating images, 173; and religious conversion, 101–04; sacred threat rituals, 168–70; scarf-giving, 174; stereotypes, 92; supporting, 102–03; undermined in cleaning duties, 66; undermined in schools, 61–62; undermined in water collection, 95–97; usurping hierarchy, 180–81, 183–84. *See also* mixed marriages; priests; religious conversion

caste system in Australia: overview, 166; caste markers removal, 187; clubs, 179–81; community events, 181–85; competing notions of community, 166–67, 168, 175–81, 183–84; and education, 186–87; hopeful expectations, 185; and neighborhoods, 186; performance of egalitarianism, 171, 176, 185; public forum on, 176; scarf-giving, 174–75; service providers intervening, 177; and Westerners, 171–72; Westpac building, 165–66

charity: Australia as generous, 20; vs. obligation, 11–12; and power, 14; refugees giving, 70; vs. rights, 17, 190, 201

Chatterjee, P., 22

Chhetris overview, 92, 94. *See also* caste system

children, 66–67, 76–77

Chimni, B.S., 6, 8

China, 38

Christianity, 99–104

Christmas, 103–04

citizenship: in Australia, 128, 199–203; in Bhutan, 40, 42, 76; and identity, 89, 194–95; and refugee

label, 200–02; vs. refugee status, 162; as value, 55–56
Citizenship Act (Bhutan) (1985), 40–41
Citizenship Act (India) (1955), 43
civil rights movement, 117
Clarke, M., 13
Cold War, 4–5
colonialism, 34–38, 115
Commonwealth Racial Discrimination Act, 117
community: BAASA, 155–56; and caste system, 98–102, 166–67; competing notions in Australia, 166–67, 168, 175–81; defined, 56; defined by "traditionalness," 198; as ethnic in Australia, 125–26; as organizing new arrivals in Australia, 136, 154; and religious conversion, 104; and Westerners, 172
compassion: and power, 140–41, 207–08; vs. rights, 202, 206, 208; withdrawn, 16, 210, 214
consent forms, 173
containment, 10, 21–22, 53–54
contraceptives, 67
contributing to society: overview, 198–200; children in rice harvests, 66; and creation of narratives, 191–95; exceptions, 162; and immigration policy, 117; as minimized, 151; and nation building, 124, 126, 127; and trauma focus, 152–53. *See also* citizenship; employment; volunteerism
Convention on the Rights of the Child, 77
Convention Relating to the Status of Refugees, 118

conversion, 99–104, 105
cows, 103–04

Darjeeling, 35–36, 37, 39
Deakin, Alfred, 116
democracy: and Australian resettlement, 128, 191–93, 196; and caste system reform, 196–97; demonstrations for, 41–42; and humanitarianism, 10, 19; in Nepal, 46; transition to, 34
demonstrations, 41–42
Department of Immigration and Border Protection (Australia), 113
depoliticization, 14–15, 17–18, 149, 163–64, 207–08
deserving. *See* worthiness
desirability, 3, 189–90, 205–06. *See also* worthiness
development projects, 8–11, 90
dignity, 151, 162
diversity, 126, 135. *See also* multiculturalism
divorce, 77–78, 79
domestic violence, 71–72, 73–74, 145
domestic work, 149–51
Dorsett, P., 21, 152
dowry, 146, 147–48
Driglam Namzha, 41
Drukpa people, 36, 37, 39–40, 41–42, 55–56, 63
Duar War, 35, 39
Duffield, M., 10
Durkheim, E., 170
Dzongkha language, 55

economics, 119
education: BAASA school, 156–58; becoming Bhutanese, 55–57; Bhutanese in Australia, 138–39; and caste system, 61–63, 95, 98,

99, 186–87; and gender, 63, 64–
  65; locals vs. refugees, 58, 60–61;
  and marriage, 99; and refugee
  status, 63; teachers, 88–89
Elizabeth, Australia, 129–30, 134
employment: and aid workers, 17,
  58, 69–70, 85; in Australia, 149–
  53, 178, 189–90; in camp schools,
  88–89; despite citizenship, 201;
  outside camps, 85, 86–88, 89, 90;
  priests, 94; Salisbury issues, 134–
  35; women in Australia, 149–51.
  See also volunteerism
equality, 65, 71, 101–02, 200–03
Eriksen, TH., 33
ethics, 28–29, 123, 143. See also
  values/norms
ethnic cleansing, 42
Ethnic Leaders Forum, 113, 198
ethnography, 22–29
etiquette, 41
Europe, 4–6
exile stories, 191–97
exotic otherness, 131

family separation, 79, 146, 202–03
family structure, 72–73, 143–44, 147.
  See also polygamy
Fanjoy, M., 21
farming, 150
Fassin, D., 16–17, 120, 146, 212
Feldman, I., 209, 211, 214
Ferguson, J., 10
fertility, 66
Fiddian-Qasmiyeh, E., 18–19
fieldwork, 22–29, 57, 64–65, 83–84;
  ethics, 28–29
Finland, 190
FitzGerald, DS., 11
follow the people, 24–25
food, 61–62, 179, 181–82, 183

Foreigners Act (India) (1946), 43
funding: and documents on
  worthiness, 15; as focusing
  on negative, 74; and morality,
  75; and suffering, 75, 152;
  and transformation, 71; and
  vulnerable people, 70, 71, 73–74

Gabiam, N., 75, 90, 209
gender: and democracy knowledge,
  191; and education, 63, 64–65;
  equality education, 71, 150; and
  public space, 144–45; roles of, 66,
  67, 72. See also men; women
Gender Focal Point, 71
ghettoization, 133
global inequalities, 9, 13
globalization, 6
Gorkhaland movement, 44
government oppression, 56

Hage, G., 21, 126
Harmony Day, 135–36
Harrell-Bond, B., 15, 17
helplessness, 15, 21, 142
Heyman, JM., 115
Hinduism: and Bhutan history, 39;
  and caste system, 91, 93, 95, 103,
  169; and charity, 70; and cows,
  103–04; marriage, 40; sacred
  threat rituals, 168–69; scarf-
  giving, 174; as strategy, 105; Teej
  in Australia, 181–85
housewarming parties, 168, 170
humanitarianism: overview, 22,
  162; Australia's history of,
  120–21, 123; campaigns, 6–7; of
  camps and suburbs, 3–4; and
  containment, 10–11; defined,
  12; and democracy, 10, 19; vs.
  development action, 9–11; four

principles of, 12; gestures vs. political change, 206–07; and global inequalities, 9, 13, 20–21; as governance, 207–13, 214; and gratification, 13; hierarchies, 140–43; history of, 12–13; and morality, 11–12, 13–14, 16; and poverty, 6; selectivity, 141; and suffering, 12–13; urgency of, 5. *See also* United Nations High Commissioner for Refugees (UNHCR); values/norms
Humanitarian Reason (Fassin), 16
Humanitarian Visa Program (Australia), 120, 121
human rights: activism, 41–42, 80, 192, 194–95; and BAASA, 156; and caste system, 61; and gender equality, 71, 80; and language banning, 156; and Refugee Convention signatories, 46; as refugee value, 10, 61; and resettlement priorities, 189; and UNHCR, 4–5
Hungary, 119
Hutchinson, M., 21, 152
Hutt, M., 37, 92
Hutus, 17–18, 89

ideal refugees, 2, 18–19, 109–10, 142–43, 149–50. *See also* performance
The Ideal Refugees (Fiddian-Qasmiyeh), 18–19
illiteracy, 173, 179–80
Immigration Restriction Act (Australia), 115, 116
imperialism, 9
Imposing Aid (Harrell-Bond), 17
India, 43–45, 47, 63
Indigenous identities, 105–07. *See also* Aboriginal people

individualism, 68
infrastructure, 58, 134–35
innocence, 15
integration: Australian policy, 124–25; Australian services, 127, 130–31; Bhutanese in Nepal, 48–49, 105–07; and caste marker removal, 187–88; and repaying Australia, 144; transforming victim roles, 188–89; and trauma in Australia, 21, 152–53. *See also* Australia resettlement; resettlement
intergenerational tensions, 27
internal policies vs. geopolitical realities, 8, 10–11
International Convention on the Elimination of All Forms of Racial Discrimination, 117
interviewees, 22–29, 83–84
intra-female violence, 72, 73–74, 145
Iraq, 16

Jupp, J., 115–16
justice: vs. benevolence, 202; and humanitarian governance, 208, 213; and refugee label, 18, 23. *See also* human rights; rights vs. charity

Kenny, P., 24

language, 41, 63, 64n1, 88, 156–58, 179–80
Lhotshampa people, 36, 37n1, 39
Limbu, B., 12
locals vs. refugees: boundaries between, 89–90; compassion for, 140–41; culture, 59–60; education, 58, 60–61; interactions spoiling identities,

95; marriage, 78; paid work, 58, 69–70, 85; resources, 57–59, 61, 127; righteousness, 60–61; and values, 62; work exploitation, 88
Lockwood-Kenny, K., 24
love marriages, 99–101, 177, 183
Lubet, S., 28

MacKenzie, C., 28
Malkki, L., 17–18, 33, 54, 89, 195, 209
Maoists, 44
Marcus, G., 24
marginalization: and depoliticization, 15; of lower castes, 178n1; of Salisbury, 130, 133, 135, 159; and welfare, 190; of women, 79. *See also* vulnerability
Marlowe, JM., 21, 152
marriage, 40, 76–80, 98–104, 143, 177, 183
Marriage Act (Bhutan) (1980), 40
material goods, 160–61
men, 63, 150–54, 191–92, 194–95
Michaels, A., 92
middle castes. *See* caste system
Migrant Resource Centre (MRC), 133, 159–61, 178–79, 198
migrants, 20–21
migration diplomacy, 42
minorities in Australia (overview), 21–22
mixed marriages, 99–101, 177, 183
model refugees, 2, 109–10, 142–43. *See also* performance
monogamy, 78, 80
morality: and development, 11; and funding, 75; and humanitarianism, 11–12, 16; improving, 64–65, 67, 75; locals vs. refugees, 60–61; and marriage

concepts, 80. *See also* values/norms
Morrison, Scott, 123–24, 127–28
Multicultural Foodies, 179
multiculturalism, 124–27, 135–36. *See also* diversity

Nah, L., 46
narratives, 191–94, 197
National Day, 55, 81
Nationality Law (Bhutan) (1958), 40–41
nation-states: absolution, 211–13; Bhutan nation building, 39; and camps as surrogates of, 11; and causes of refugees, 8, 11; and community, 56; and ethnicity, 125; Nepal borders, 47; responsibilities and worthiness, 10; responsibility shift, 68
Nelson, A., 37n1
Nepal: and cows/beef, 103; Gorkhaland movement, 44; health of population, 57; history of refugees, 45–47; negotiating refugee flow, 42, 46, 47; refugees as resettled from, 25. *See also* camps
Nepalese: caste system blame, 65; history of, 36–37, 38–40, 44; self-identity of, 37n1; working in camps, 58, 69–70. *See also* locals vs. refugees
Netherlands, 190
Neumann, K., 119
1967 Protocol Relating to the Status of Refugees, 5, 46
nongovernmental organizations (NGOS), 214
norms. *See* values/norms

Index 245

Ogura, K., 103
Ong, A., 54, 68, 161
Operation Sovereign Borders, 20
oppression, 56
Orientalism, 131

Palestine, 75, 90
participation, 23-24, 67, 138, 174-80, 185, 191. *See also* contributing to society
passivity, 190
pedestrian crossings, 201
Penguin Club, 179-81
performance: overview, 3; affirmative Orientalism, 131; changing nationality, 16; of culture, 138-39; of egalitarianism, 171, 176, 185; and group representation, 84; and masking divisions, 84; "refugee-ness," 90-91, 143, 197; and refugee status, 2; regulating images, 2, 83, 97; women as representatives, 142-43. *See also* transformation
personal empowerment, 68
Phillips, J., 151
political activism, 7-8, 74, 191-92, 194-95
political prisoners, 41-42
pollution. *See* purity/pollution
polygamy, 77-80, 146-47, 148
poverty: and globalization, 6; and humanitarianism, 6; refugees helping, 70; in Salisbury, 134-35; villages close to Belgandi camps, 58
power imbalances: overview, 204; and control, 154; despite citizenship, 201-02; dominant and minority groups in Australia, 21-22, 201, 204; encouraged through colonialism, 37; and humanitarianism, 20-21, 162; leaders in camps, 63-64; refugees/aid workers, 17; researchers, 28-29. *See also* caste system
pregnancy, 27-28
priests, 92, 94, 103, 139
privileges, 69
protests, 41-42
public transportation, 86-87
Purity and Exile (Malkki), 18
purity/pollution: and food, 171-72, 179, 181-82, 183; and marriage, 98-99; and sacred threat ceremony, 168-69; and *tikkas*, 165; and water, 95; and Westerners, 171-72

racism, 116-17, 125, 134, 159. *See also* white supremacy
rape, 42
Redfield, P., 212
Refugee Convention, 46
refugees: vs. asylum seekers, 113-15, 121-22; and Australia economy, 119; Bhutan government attitude, 56; defined, 4; demonstrating genuineness, 89; feminization of image, 142-43, 149-50; as helpless, 15, 21, 142; history of, 4-10; idealized, 109-10; and identification, 37n1; vs. migrants, 20-21; number migrating into Europe, 214; as pawns, 42, 45; perception of, 6, 7-8, 16, 89-90; ranking, 15, 19; resource entitlement, 57; root causes of, 8, 11; as social category, 207-08; as specimens, 210; as

successful, 90; as threat, 124, 206, 214–15; as victims, 6–7, 12, 13–17, 90, 122. *See also* asylum seekers; Australia resettlement; camps; caste system; caste system in Australia; locals vs. refugees; performance; resettlement; resources; trauma
refugee status: and caste system, 166; vs. citizenship, 162; defined, 4; despite citizenship, 200–02; and education, 63; and expected responses, 1–2; and gender, 150; labels and recognition, 18; and political status, 14; as privilege, 14; and recipient role, 199; requirements of, 194; as undermining, 151–53, 162; universalizing, 5
refugee warriors, 7–8
regulating images: and caste system, 173; to outsiders, 173; performance, 2, 83, 97; for resettlement, 57; and resources, 85; Syrians posing as Kurds, 16; women as representatives, 142–43
regulating movement, 53–54
reinvention, 68–69
religion, 98. *See also specific religions*
religious conversion, 99–104, 105
repatriation, 5, 47, 55–56
research, 22–29, 57, 64–65, 83–84
resettlement: overview, 107–10; Bhutanese and West, 2; and causes of refugees, 8; during Cold War, 5; countries chosen, 108–09; vs. development projects, 8–9, 11; and polygamy, 80; and regulating images, 57; vs. repatriation, 5; and scrutiny, 77; and working refugees, 88; and worthiness, 90–91. *See also* Australia resettlement
resources: Australia overview, 130–31; and development, 9; refugees vs. asylum seekers, 114; refugees vs. hosts, 57–59, 61, 127; and regulating images, 85
responsibility, 68
rice cookers, 160–61
rice cultivation, 86–88
righteousness: Bhutanese vs. Nepalese, 60–61, 85; and competing notions of community, 83; and dowry, 148; and resettlement, 75; of Salisbury, 155
rights vs. charity, 17, 190, 201
Rose, N., 68
Royal Bhutan Army, 42
Royal Bhutan Police, 42

sacred threat rituals, 168–70
Sahrawi camps, 18–19
Salisbury, Australia (overview), 128–30, 133–35, 139–40, 154–55, 158–59, 186. *See also* Australia resettlement
Salter, M.B., 47
Sanischare camp, 49
Saul, B., 42
scarf-giving, 174–75
schools. *See* education
Schuster, L., 120
security, 53–54, 135
self-sufficiency, 68, 91
Serbia, 14
serfdom, 36
service providers in Australia: overview, 128–29, 131; and caste system, 168, 171, 172–77, 184–85;

Index 247

Salisbury vs. Adelaide, 155, 159.
  See also Migrant Resource Centre
Sharchopas people, 107
Shneiderman, S., 103
Sikkim, 38
Smuggling of Migrants by Land, Sea
  and Air protocol, 120
soccer, 198–99
social exclusion, 190–91
social welfare, 127, 189–90
Squire, V., 120
Stam, K., 37n1
Stein, BN., 6
stereotypes, 92, 138
Sudan, 21, 152
suffering: vs. acting, 15; and funding,
  75, 152; and humanitarianism,
  12–13, 14–15; narratives
  supporting, 192, 194
suicides, 107
surveillance, 135
surveys, 64–65
Survivors of Trauma and Torture
  Rehabilitation Services
  (STTARS), 130
Symons, J., 115
Syria, 16, 205, 206

Tanzania, 17–18, 89
taxes, 38, 39
teachers, 88–89
Technical and Further Education
  (TAFE) (Australia), 127
Teej celebration, 181–85
10M, 76, 108, 109
terrorism, 206, 214
Thomas, DA., 13
Tibet, 38, 47
Ticktin, M., 151, 214
tikkas, 165, 169, 187, 196
torture, 139. *See also* trauma

transformation: in camps
  (overview), 209–10; and
  funding, 71; illustrations
  of, 66, 67, 69, 70–71, 73–74,
  77–80; and interventions, 10,
  71–72, 73–74, 77, 79–80; and
  marginalization, 78–79; morally
  righteous values, 64–65, 67;
  people vs. conditions, 206–07;
  vs. physical containment, 53–54;
  and precarious status, 20; for
  protection, 80; reinvention,
  68–69; and values/norms,
  54–55, 64–65, 67, 69. *See also*
  performance
transportation, 86–87
trauma: as dominant identity
  feature, 152–53; and equality,
  194–95; and worthiness, 21,
  122–23, 189
Treaty of Friendship (India and
  Bhutan) (1949), 43
Treaty of Peace and Friendship
  (India and Nepal) (1950), 45
Tsourapas, G., 33–34

Uganda, 17
United Liberation Front of Assam,
  44
United Nations Convention against
  Transnational Organized Crime,
  120
United Nations Convention
  Relating to the Status of
  Refugees, 4
United Nations Declaration on the
  Rights of Indigenous Peoples,
  106
United Nations Development
  Programme (UNDP), 9

United Nations High Commissioner for Refugees (UNHCR): and BAASA, 156; campaigns, 7; documents on worthiness, 15; end of resettlement involvement, 213; history of, 4–10; influencing Australia, 189; and marriage concepts, 76; and model refugees, 2; and Nepalese host population, 58; presence in Beldangi camps, 53–54; and refugees in Nepal, 47–49, 48; refugees thanking, 27; as supra-government, 145–46. *See also* camps; humanitarianism
United States, 108, 120, 189
Universal Declaration of Human Rights, 117, 156

values/norms: citizenship, 55–56; as cultural problems, 134; and gatekeeping, 70–71; and identity, 148–49; internalizing, 67, 69, 70–71, 75, 77, 98; and refugees as "underdeveloped," 11; in schools, 61–62; and transformation, 54–55, 64–65, 67, 69. *See also* caste system; dowry; performance; polygamy; transformation
Vietnam, 119–20
violence, 71–72, 73–74, 145
virginity, 76
volunteerism, 69, 70–71, 85, 86
vulnerability: and desirability, 189–90, 205–06; and funding, 70, 71, 73–74; and ghettoization, 133. *See also* women

Wangchuck, Ugyen, 36
War-Time Refugees Removal Act (Australia), 118

water collection, 95–97
Watson, SD., 20
wells, 95–96
Western countries, 42
Westpac House, 165
West resettlement (overview), 2, 5, 11
White Australia Policy, 115, 116
white supremacy, 116–17, 126–27. *See also* racism
Whitlam, Gough, 117
Wolf, ER., 33
women: and BAASA, 159; as camp leaders, 63–64; cooking rice, 160; and education, 63, 64–65; emancipation of, 144–45, 179–80; in family structure, 72–73, 143–44; and fertility, 66; focus on in Australia, 141–45, 149–50; as under guardianship, 144; narratives of, 193; and Neikirk's pregnancy, 27–28; and polygamy, 80; violence by women, 72, 73–74, 145; water collection, 95–96; working in Australia, 149–51. *See also* dowry; polygamy
Woodville, Australia, 186
working class, 129
worthiness: overview, 13–15; and asylum avenues, 114–15; and "bare humanity," 7; vs. development projects, 90; and helplessness, 15; and living conditions, 75; narratives supporting, 194; and political activism, 202; as precarious, 201–02; refugees vs. migrants, 20–21; and resettlement, 90–91; Sahrawi camps, 18–19; of Serbian refugees, 14; and state responsibilities, 10; after

Index 249

terrorism attacks, 214–15; and trauma, 21, 122–23, 189; and visas, 202. *See also* asylum seekers; desirability; values/norms

yoga, 137
Yugoslavia, 14

Zetter, R., 18, 20

Other Titles from University of Alberta Press

Contemporary Indigenous Cosmologies and Pragmatics
Edited by FRANÇOISE DUSSART and
SYLVIE POIRIER
Thirteen contributors examine Indigenous peoples' negotiations with different cosmologies in today's globalized world.

Countering Displacements
The Creativity and Resilience of Indigenous and Refugee-ed Peoples
Edited by DANIEL COLEMAN, ERIN GOHEEN
GLANVILLE, WAFAA HASAN, and
AGNES KRAMER-HAMSTRA
Collection of essays forges compelling linkages between cultural experiences of refugees and Indigenous peoples worldwide.

Narratives of Citizenship
Indigenous and Diasporic Peoples
Unsettle the Nation-State
Edited by ALOYS N.M. FLEISCHMANN,
NANCY VAN STYVENDALE, and
CODY MCCARROLL
Thirteen essays examine literature, film, music, etc. to conceptualize citizenship as a narrative construct.

More information at uap.ualberta.ca